PRAIRIE-POINT PIZZA

QUILTS WITH
NEW DIMENSION

KAREN SIEVERT

Martingale®
& COMPANY

Prairie-Point Pizzazz:
Quilts with New Dimension
© 2011 by Karen Sievert

That Patchwork Place® is an imprint of
Martingale & Company®.

Martingale & Company
19021 120th Ave. NE, Ste. 102
Bothell, WA 98011-9511 USA
www.martingale-pub.com

CREDITS

President & CEO ▸ **Tom Wierzbicki**

Editor in Chief ▸ **Mary V. Green**

Managing Editor ▸ **Karen Costello Soltys**

Design Director ▸ **Stan Green**

Technical Editor ▸ **Ellen Pahl**

Copy Editor ▸ **Sheila Chapman Ryan**

Production Manager ▸ **Regina Girard**

Illustrator ▸ **Robin Strobel**

Cover & Text Designer ▸ **Stan Green**

Photographer ▸ **Brent Kane**

Printed in China
16 15 14 13 12 11   8 7 6 5 4 3 2 1

**Library of Congress Cataloging-in-Publication Data is
available upon request.**

ISBN: 978-1-60468-063-8

Mission Statement
Dedicated to providing quality products
and service to inspire creativity.

DEDICATION

With love and heartfelt gratitude to my mother, Marlene Johnson, and to the memory of my mother-in-law, Jennie Miles. It wasn't until I became a mother that I fully understood all the things a mother *does*, all the things a mother *is*, and all the ways a mother *loves*. Thank you both for being such a shining example, for teaching me, and for loving me. I can't begin to tell you how very much you mean to me!

ACKNOWLEDGMENTS

Love is a circle—composed of family, friends, and supporters. It's a community of key people who see the best in you, believe in you, and care about you without condition. It's what you give, and hopefully, what you get back.

Vince, my husband, I love you and I thank you. Sometimes the simplest words are the best. I hope you know how much your love and support have meant to me over the last 26 years. *You* are the rock star!

Wayne, Shannah, and Travis, my children—I'm so glad to be your mom. You all are the greatest! Thanks for helping me, making prairie points, assisting at trunk shows, and even commenting on projects. What would I do without you?

Thank you to my longtime friend and sidekick, Phyllis Marshall. I met you at the first class I taught and we've been the best of friends ever since. Every woman should have a friend like you!

My sisters, Barbara, Kathy, and Lori, are the best—God made us sisters, but life made us friends.

Janice Streeter, little do you know how inspirational you are. Your "Autumn Leaves" quilt struck a chord in me; it made me want to create, motivated me to discover "my own way," and set me on an incredible journey.

Rae Ann Voelkner, thanks for the beautiful hand quilting. I appreciate not only your friendship, but also your time and talent.

Vicki Maloney, I was so impressed by your beautiful machine quilting from the first moment I saw it. I'm very grateful for your workmanship on the basket sampler project. Thank you ever so much. Your quilting is some of the best I've seen.

Thanks to the Haymarket Chapter of Quilters Unlimited for your love and support. For a small chapter, it sure has a whole bunch of very talented women. I'm so very thankful for your friendship.

Cotton Club Quilters, I had the *best* time at your retreat. And you all were instrumental in discovering even more ways that the "Color, Color, All Around" blocks could be laid out. You are a great group of women and a phenomenal group of quilters.

Thanks to Martingale & Company for taking a chance on me. Little girl, big dream . . . that you all helped come true!

Quilters, quilters, everywhere, thank you! My dad used to tell me, "There's no such thing as a stranger, just friends we haven't met yet." That's so true! Throughout my travels, I feel I've been blessed by both your friendship and your acceptance.

CONTENTS

INTRODUCTION

Have you ever wanted to add another element, a little something extra, to your quilts? Have you wanted them to stand out, have a little more "pop," or be unique? Well, that's what this book is all about. *Prairie-Point Pizzazz* introduces a new and exciting technique for incorporating prairie points into any half- or quarter-square triangle, immediately adding dimension to traditional quilt blocks.

I've always referred to quiltmaking as a journey, a trip that teaches us much. We experiment with color, we learn and try new techniques, and somehow, we grow as quilters. My journey began in 1997, when my sister dragged me into a quilt shop. But more than that, my journey progressed as I went to shows and saw what quilters were capable of—the unbelievable creativity of so many people!

In 1998, I went to our local quilt guild's show. There I saw a quilt titled "Autumn Leaves" by Janice Streeter, an incredibly talented master quilter. Janice had used prairie points on bias strips to make her leaves dimensional. To this day, I remember it as one of the most beautiful quilts I've ever seen.

Well, needless to say, visions of sugarplums, or rather prairie points, were dancing through my head! But, I needed an *easier* way to incorporate prairie points into my quilts because bias strips just didn't work for me (they curve!). I tried and tried, as many of us do, and I finally failed my way to success. After many years of experimentation, figuring out what *didn't* work, I was finally able to figure out a way that *did* work.

And what a discovery that was. As you read through this book and use the technique, I think you too will be amazed. It's so incredibly simple! There's no need to redesign a block—just use the easy technique and placement guide to add the prairie points to any half- or quarter-square triangle. You can purchase an acrylic placement guide, but there are also instructions for easily making your own if you prefer.

So, let's get started. As you travel through the book, there are a few stops you'll want to make. First, there's a must-read section about prairie points and how to use them. Next, you can meander through some really great patterns. If you're a mother or grandmother, then the "Peekaboo" pattern (page 12) was designed just for you. If you're like me, you'll be enchanted with the dimensional butterflies in "Flutterbyes in My Garden" (page 16). And everyone loves the endless possibilities of "Color, Color, All Around" (page 29). Finally, the "Quiltmaking Basics" (page 72) section offers tips and suggestions to make your quilting experience easier.

I hope this book will leave you with visions of prairie points dancing through your head! I'm sure you'll enjoy adding pizzazz to your quilts with prairie points just as much as I have. The possibilities are infinite, and as much as I'm excited about these patterns, I'm just as excited about the ones still to be realized.

May God bless you, your families, and yes, your quilts.

~Karen

PRAIRIE POINTS

Here it is . . . the heart and soul of this book. Since all of the patterns have been designed to incorporate prairie points, I'll start with what they are, how they're made, and how they can be inserted into your patchwork. Additionally, I'll share some tips for choosing fabrics for prairie points as well as a foolproof way to determine what size of prairie point you'll need to adapt this technique to blocks that aren't in this book. There are so many ways to take a block and make it new and exciting, and I believe that once you see how incredibly easy this is, you'll be seeing endless possibilities too.

So, where have you seen prairie points? Typically, when you see prairie points in quilts, they're a decorative accent in the border or binding, and not usually in the body of the quilt. Prairie points actually started out as a trim on garments in the mid-1800s, with their popularity increasing and peaking around 1930. However, they weren't called prairie points until around 1960.

If you were to look up the definition of *prairie point* in a quilt book, it would read something like, "a folded fabric triangle used as a quilt border or embellishment." That's a wonderful use for prairie points, but as you page through this book, you'll see that prairie points inserted into your patchwork can truly change the landscape of your quilts. Using them in the body of the quilt adds a whole new layer of interest.

The beauty of it is, prairie points are incredibly easy to make. Once made, all you do is fuse them to half- or quarter-square triangles, and the rest is just regular piecing. If you know how to piece, then you can do this! So, let's get started; let me show you how truly simple and easy this is.

FABRIC SELECTION

Even if you're a pro at selecting fabrics for your quilts, you may find some helpful information here on selecting fabrics for the prairie points, as well as on fabric preparation to make handling those prairie points easier.

I absolutely love picking out fabric for a new project. Every new quilt feels like a new beginning. Yet, there was a time when I agonized over this basic function of quilting. It was so intimidating. I believe all of us go through this at one time or another, but there are some tricks that can help make picking out your fabrics a much easier process. I have a habit of walking into a shop and spotting a fabric that's to die for. For me, it's always a multicolored print that somehow just sings to my soul. Usually I'll snatch it up, hug it to my chest, and walk around the shop with a death grip on it, afraid someone else might want it.

But here's the thing: once you find that one beautiful fabric, picking out all your other fabrics becomes easy. Look at the selvages and you'll see dots of all the colors used in the printing process. That's a great way to choose coordinating colors. After a while, you won't need to look at the selvages; you'll instinctively know what works.

When you find your to-die-for print, you can easily pick out coordinating fabrics.

In addition, you'll want to look at the pattern, size, and scale of your prints. Large-scale prints work well in patterns that have large pieces, but aren't that effective when cut into small pieces or folded into prairie points. Small-scale prints generally work better in patterns that have lots of small pieces. The prairie points throughout the book are made from squares of fabric that have been folded into quarters. You'll want to look at how that fabric shows when folded. I tend to use small-scale prints or tone-on-tone fabrics for prairie points.

The style of the fabric can also influence your choices, and there are many styles out there—from '30s reproduction prints to bright and bold novelty prints, from Asian prints with metallic hints to themed prints. While all are beautiful, they are usually best paired with other fabrics of a similar nature.

Remember, fabric selection should be fun. Keep all the above in mind, but don't be afraid to experiment; every quilter has had both success and failure with fabric choices. It's how you learn and grow and how you find your own quilting personality. It's one of my favorite things about quilting—even if we all start with the same quilt pattern, we end up with entirely different results based on our fabric and color choices.

MAKING PRAIRIE POINTS

Prairie points are nothing more than folded squares of fabric. There are a couple of different ways to make prairie points, but in all of the patterns in this book, I have used just one technique. When pressing prairie points, I recommend using steam. This is the only time I use steam—it makes the folded edge of the prairie point much crisper. (The only problem is that it often fogs up my glasses!)

To make the prairie points in this book, you'll need just a few supplies:
- Steam iron and ironing board
- Square of fabric (as directed in each project)
- ¼"-wide Steam-A-Seam 2 (optional)

1. At the ironing board, fold the square of fabric in half, wrong sides together, to make a rectangle. Press with the iron on the steam setting.

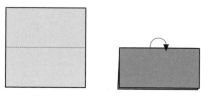

2. Bring the left and right folded corners toward the center of the raw edges of fabric as shown. Press.

3. This is optional, but I like to place a small piece of ¼" Steam-A-Seam 2 (½" to 1" long) along the bottom of the prairie point to fuse the folded corners in place. It helps to keep the prairie point from popping open and avoids sewing challenges later.

¼" Steam-A-Seam 2   Corners are fused in place.

HANDY HELPER

Recently I came across a great gadget called the Prairie Pointer. This stainless steel pressing tool aids in creating sharp-edged prairie points and helps you line up the center fold perfectly. While all the prairie points in this book were made without the tool, I can see why quilters are swearing by it. Additionally, you can use it to determine the size of square needed for a prairie point. See "Making Prairie Points for Any Block" (page 11).

FUSING PRAIRIE POINTS

After making the prairie points, you'll fuse them to the appropriate patchwork piece, usually a half- or quarter-square triangle. To make sure that the prairie points are placed uniformly and centered within the triangles, you'll need a placement guide.

When I first started designing patterns with prairie points, I used heat-resistant template plastic to make placement guides. Since then, I've developed an acrylic version of the template, which I absolutely love because of its weight and ease of use. You can easily make your own placement guide, but if you'd like an acrylic template, it's available solely through my website. (See page 79).

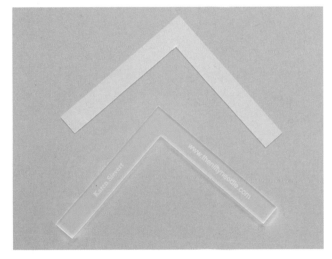

Placement guides made from template plastic (*top*) and acrylic (*bottom*)

What You'll Need

- Prepared prairie points
- Half- or quarter-square triangles as directed in the project instructions
- ¼"-wide Steam-A-Seam 2
- Placement guide or purchased Acrylic Placement Guide
- Steam iron and ironing board

Placing and Fusing Prairie Points

1. To make your own placement guide, trace the pattern (page 10) onto heat-resistant template plastic, regular template plastic, or other sturdy material and cut out exactly on the drawn lines. This guide can be used for triangles and prairie points of any size.

2. At the ironing board, position the placement guide on the fabric triangle so that its edges are evenly aligned with the short edges of the triangle. Place a piece of ¼" Steam-A-Seam 2 along the long edge of the triangle as close to the raw edge as possible.

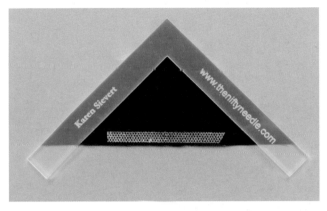

Positon placement guide and Steam-a-Seam 2.

3. Place the prairie point in the open space, snugging it up to the placement guide without moving the placement guide.

Position the prairie point.

4. Move the placement guide out of the way without moving the prairie point. Fuse the prairie point into place using a steam iron. This will typically take about three to five seconds, but since all irons are different, you'll need to determine the amount of time to keep your iron in place for effective bonding.

You may want to keep a pressing cloth or piece of parchment paper handy in case you need to protect your iron or ironing surface from excess fusible web.

Move placement guide and fuse.

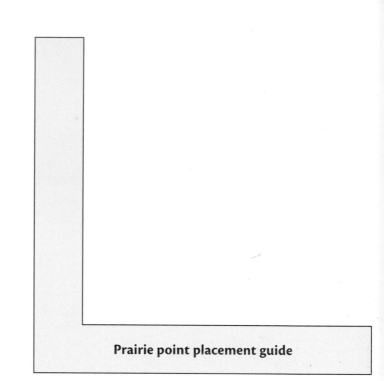

Prairie point placement guide

The beauty of the ¼" Steam-A-Seam 2 is this: it's one of the best and quickest bonding agents, and when fused, it dries clear. When placed along the raw edge, covered with the prairie point, and then sewn, it is totally *invisible* in the ¼"-wide seam allowance. No one will ever know it's there. (Unless you tell them. But why would you?) It's like magic—magic that eliminates pinning and shifting, magic that you can sew through, and magic that makes this whole process so simple. See, I told you it was easy!

FUSING IN A NUTSHELL

To fuse prairie points, you only need to remember three simple steps: *place, move,* and *fuse.*

I've always been a visual learner. For others like me, there's a short video on my website demonstrating how to make prairie points, how to fuse them, and even how to determine square size for them. It only takes a few minutes to demonstrate, so that should tell you how easy this process really is! Be sure to go to www.theniftyneedle.com to view the video.

ADDITIONAL TIPS AND TRICKS

In this section are a few additional tidbits of information that I've learned from drafting and making prairie-point quilts.

Prewashing, drying, and pressing your fabrics with a good quality spray starch is important for making prairie points. By adding sizing, you'll have a nice crisp fabric that yields sharp edges and points when you fold and press the prairie points.

Because the folded squares of fabric aren't always exact, the raw edges of the prairie point may sometimes extend below the raw edges of your triangle. If that happens, trim the prairie point after fusing so that the raw edges all align. It's very important that you don't cut the half- or quarter-square triangle, since those are cut precisely for the pieced block.

Traditionally, when prairie points are used in quilts, the center folded edges face outward and are visible. In all the patterns in this book, I've placed the folded edges facing the quilt top. This hides any imperfections in the folding process, and there are no folds on the exterior to get caught in. Whether you decide to place them with the fold visible or not, be consistent to create a cohesive, rather than haphazard look.

Fold facing up
(visible)

Fold facing down
(invisible)

You'll find that there's no difference piecing blocks with prairie points than blocks without them. Today's sewing machines are perfectly capable of handling the little bit of extra bulk. Once when teaching, I came across a machine with a presser foot that tended to ride along the edge of the prairie point instead of riding over it. Sewing at a steady, moderate pace solved the problem. That gives *you* control over the fabric feed as opposed to the machine.

MAKING PRAIRIE POINTS FOR ANY BLOCK

If you want to insert prairie points into patterns not in this book, all you need to do is determine the square size needed for the prairie point. The block pieces don't change; you'll just add prairie points to them. There are several ways to do this.

As an example, let's say that you want to make a star block and add prairie points to the star points. In the block pattern, the star points are created from 4½" squares that have been cut twice diagonally to make quarter-square triangles.

Using the Prairie Pointer:

1. Cut a 4½" square from scrap fabric, cut it into quarters diagonally, and place one of the triangles on a flat surface. (You can also use a piece of paper for this, rather than fabric.)
2. Position the placement guide on the triangle just as you would for fusing, aligning the edges evenly.
3. Snug the Prairie Pointer tool into the remaining space. Move the placement guide away and observe where the fabric line meets the tool. In this example, you would need a 3" square of fabric to make the appropriate-size prairie point.

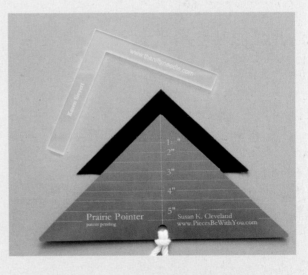

Measuring:

Follow steps 1 and 2 above; then simply measure the distance along the bottom of the triangle between the inner edges of the placement guide as shown at right.

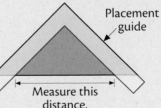

Placement guide

Measure this distance.

Using Formulas:

1. For quarter-square triangles, subtract 1.414" from the original cut square size, and then round down to the nearest ⅛" (.125"). In the example above, it would be 4.5" − 1.414" = 3.086", or 3".
2. For half-square triangles, multiply the original cut square by 1.414", subtract 1.414", and round down to the nearest ⅛". For a half-square triangle cut from a 3⅞" square, it would be 3.875" x 1.414" = 5.479"; 5.479" − 1.414" = 4.065", or 4".

PEEKABOO

There's something truly appealing about this very simple quilt. I think it's the hidden treasures—fused fabric images—that little ones will find under the prairie-point flaps (see the close-up on page 14). Every time I see this quilt, I come up with more ideas. Imagine if, under the prairie points, there were handwritten notes about why you felt so fortunate to have the recipient in your life. The possibilities are endless. Where will your imagination take you?

Finished quilt: 40½" x 50½"
Finished block: 10" x 10"

MATERIALS

Yardage is based on 42"-wide fabric unless otherwise indicated.

- ▶ 1⅜ yards of blue print for border and binding
- ▶ ½ yard *each* of light pink, light yellow, light blue, and light green fabrics for blocks
- ▶ ⅓ yard *each* of medium pink, medium yellow, medium blue, and medium green fabrics for prairie points
- ▶ ¼ yard *each* of dark pink, dark yellow, dark blue, and dark green fabrics for blocks
- ▶ 2⅔ yards of fabric for backing
- ▶ 48" x 58" piece of batting
- ▶ 48 print motifs, approximately 1½" x 3" each, for fused appliqués
- ▶ ½ yard of 18"-wide fusible web (I used HeatnBond Ultrahold)
- ▶ 9 yards of ¼"-wide Steam-A-Seam 2
- ▶ Placement guide (see page 9)

CUTTING

All measurements include ¼"-wide seam allowances.

From *each* of the light fabrics, cut:
- 1 strip, 5⅞" x 42"; crosscut into 6 squares, 5⅞" x 5⅞". Cut each square in half diagonally to yield 12 triangles (48 total).
- 1 strip, 6¼" x 42"; crosscut into 3 squares, 6¼" x 6¼". Cut each square into quarters diagonally to yield 12 triangles (48 total).

From *each* of the medium fabrics, cut:
- 2 strips, 4¾" x 42"; crosscut into 12 squares, 4¾" x 4¾" (48 total)

From *each* of the dark fabrics, cut:
- 1 strip, 6¼" x 42"; crosscut into 3 squares, 6¼" x 6¼". Cut each square into quarters diagonally to yield 12 triangles (48 total).

From the blue print, cut:
- 5 strips, 5½" x 42"
- 5 strips, 2" x 42"

PIECING THE BLOCKS

1. Referring to "Making Prairie Points" (page 8), make prairie points from *each* of the medium 4¾" squares for a total of 48.
2. Referring to "Fusing Prairie Points" (page 9), fuse the medium pink prairie points to the dark pink 6¼" triangles. Make 12. Repeat for the other three colors to make a total of 48 fused prairie points.

Pieced and quilted by Karen Sievert

PEEKABOO

3. Sew a light pink 6¼" triangle to a medium pink prairie-point unit. Press the seam allowances toward the light pink triangles. Make 12 units. Trim the dog-ears.

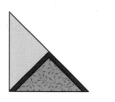

Make 12.

4. Sew a light pink 5⅞" triangle to the unit from step 3 as shown. Press the seam allowances toward the large pink triangle. Make 12. Trim the dog-ears.

Make 12.

5. Lay out four units from step 4 into two horizontal rows. Sew each row together. Press the seam allowances toward the large pink triangle. Sew the rows together to complete the block. Press the seam allowances open. Make three blocks.

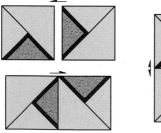 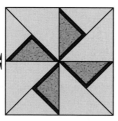

Make 3.

6. Repeat steps 3–5 to make three Pinwheel blocks in each of the other three colors.

7. To fuse the motifs under the prairie points, follow the manufacturer's instructions. Adhere fusible web to the wrong side of the motif fabric. Cut out the motifs, place them under the prairie points, and fuse in place to create the peekaboo effect.

FINDING AND FUSING MOTIFS

I made a see-through plastic template the same size as the prairie points to ensure that the motifs would fit under the prairie points without showing. In fact, this is a great idea for when you're shopping for motif fabric. Take the template along so you can be sure that the motifs will fit before you buy the fabric!

ASSEMBLING THE QUILT TOP

1. Referring to the quilt assembly diagram below, lay out the blocks in four rows of three blocks each, alternating the colors in a manner that's pleasing to you. Sew the blocks in each row together. Press the seam allowances in opposite directions from row to row. Sew the rows together. Press the seam allowances in one direction.
2. Sew the 5½" x 42" blue print strips together end to end to make one long strip.
3. Referring to "Borders" (page 74), sew the 5½"-wide borders to the quilt top using the butted-corners technique.

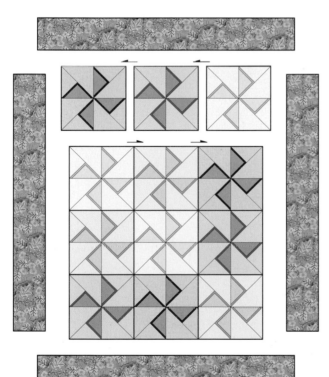

QUILTING AND BINDING

1. Refer to "Preparing the Quilt Sandwich" (page 76) to layer the quilt top, batting, and backing.
2. Hand or machine quilt as desired.
3. Referring to "Binding" (page 77), use the 2"-wide blue print strips to bind the edges of your quilt.

QUILTING SUGGESTION

Because I made this quilt with a child in mind, I kept the quilting super simple. A ¼" outline of all the non-prairie-point patchwork, stitching in the ditch around the quilt top and border, and a free-hand feather in the border made quilting this quilt quick and easy! I used Quilters Dream Puff batting to give it a little extra loft and to make it more comfortable for a baby to lie on while peeking under all those prairie points to see what's hiding there. Peekaboo, little one!

FLUTTERBYES IN MY GARDEN

When my kids were young, merely toddlers, there were certain words they just couldn't seem to say correctly. Instead of "butterflies," they'd say "flutterbyes" in a singsong voice, hence the name of this quilt, "Flutterbyes in My Garden." I love the way the prairie points give wings to the butterflies, enchanting us as they dance across the quilt.

Finished quilt: 50" x 50"
Finished block: 8" x 8"

MATERIALS

Yardage is based on 42"-wide fabric.

- 2 yards of border print with at least 4 pattern repeats*
- 1⅔ yards of batik print for blocks, prairie points, inner border, and binding
- ⅞ yard of dark purple print for blocks and prairie points
- ⅝ yard of beige print for blocks
- ⅜ yard of floral print for blocks
- 3¼ yards of fabric for backing
- 58" x 58" piece of batting
- 6 yards of ¼"-wide Steam-A-Seam 2
- Placement guide (see page 9)

This yardage allows for mitered borders. If you prefer to add butted borders, 1½ yards is enough.

CUTTING

All measurements include ¼"-wide seam allowances.

From the dark purple print, cut:
- 1 strip, 5¼" x 42"; crosscut into 6 squares, 5¼" x 5¼". Cut each square into quarters diagonally to yield 24 triangles.
- 3 strips, 4⅞" x 42"; crosscut into 22 squares, 4⅞" x 4⅞". Cut each square in half diagonally to yield 44 triangles.
- 2 strips, 2⅝" x 42"; crosscut into 32 squares, 2⅝" x 2⅝"

From the batik print, cut:
- 3 strips, 5½" x 42"; crosscut into 16 squares, 5½" x 5½"
- 2 strips, 5¼" x 42"; crosscut into 10 squares, 5¼" x 5¼". Cut each square into quarters diagonally to yield 40 triangles.
- 2 strips, 2⅞" x 42"; crosscut into 16 squares, 2⅞" x 2⅞". Cut each square in half diagonally to yield 32 triangles.
- 4 strips, 1½" x 42"
- 6 strips, 2" x 42"

Continued on page 18

Pieced and quilted by Karen Sievert

From the beige print, cut:

- 1 strip, 4⅞" x 42"; crosscut into 8 squares, 4⅞" x 4⅞". Cut each square in half diagonally to yield 16 triangles.

- 2 strips, 2⅞" x 42"; crosscut into 16 squares, 2⅞" x 2⅞". Cut each square in half diagonally to yield 32 triangles.

- 2 strips, 2½" x 42"; crosscut into 32 squares, 2½" x 2½"

From the floral print, cut:

- 1 strip, 5¼" x 42"; crosscut into:
    1 square, 5¼" x 5¼"; cut into quarters diagonally to yield 4 triangles
    4 squares, 4⅞" x 4⅞"; cut each square in half diagonally to yield 8 triangles

- 1 strip, 4½" x 42"; crosscut into 5 squares, 4½" x 4½"

From the border-print fabric, cut on the *lengthwise* **grain:**

- 4 strips, 7¼" x 72"

CUTTING BORDER PRINTS

Border-print fabrics have multiple repeats of a striped design that runs lengthwise along the fabric. They can make such a statement as a frame for your patchwork, but they look best when you miter the border corners. Cut the borders extra long so you can move them along the side of the quilt to see the most advantageous place for the miter. Then mark and pin the exact measurement as explained in "Borders" (page 74).

PIECING THE BUTTERFLY BLOCKS

These blocks are a variation of Old Maid's Puzzle. With the addition of prairie points, I've called them Butterfly blocks.

1. Referring to "Making Prairie Points" (page 8), make 16 prairie points from the batik print 5½" squares and 32 prairie points from the dark purple 2⅝" squares.

2. Referring to "Fusing Prairie Points" (page 9), fuse the 16 batik prairie points to 16 of the dark purple 4⅞" triangles. Fuse the 32 dark purple prairie points to the 32 batik 2⅞" triangles.

3. Sew a beige print 4⅞" triangle to the large batik prairie-point unit along the long edge. Press the seam allowances toward the beige print triangle. Make 16 units. Trim the dog-ears.

Make 16.

4. Sew a beige print 2⅞" triangle to a dark purple prairie-point unit along the long edge. Press the seam allowances toward the beige print triangle. Make 32 units. Trim the dog-ears.

Make 32.

5. Sew a beige print 2½" square to the right side of each unit from step 4, ensuring that the prairie point is pointing to the bottom-right corner. Press the seam allowances toward the beige square.

Make 32.

6. Lay out two units from step 5 and sew together. Press the seam allowances toward the top unit. Make 16 units.

Make 16.

7. Lay out two units from step 6 and two large prairie-point units from step 3 into two horizontal rows. Sew each row together. Press the seam allowances toward the large prairie points. Sew the rows together to complete the block. Press the seam allowances open. Make a total of eight blocks.

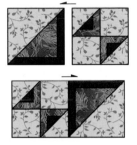 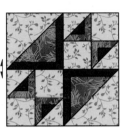

Make 8.

PIECING THE SQUARE-IN-A-SQUARE BLOCKS

1. Sew batik 5¼" triangles to opposite sides of a floral 4½" square. Press the seam allowances toward the batik triangles. Repeat for the remaining two sides of the square. Make a total of five units. Trim the dog-ears.

Make 5.

EASY MATCHING

When making Square-in-a-Square blocks, fold the square and triangle in half and finger-press to mark the center. Then when getting ready to sew, match up the centers. This will keep the blocks nice and square.

2. Sew dark purple 4⅞" triangles to opposite sides of the unit from step 1. Press the seam allowances toward the dark purple triangles. Repeat for the remaining two sides of the unit, again pressing toward the dark purple triangles. Make a total of five blocks. Trim the dog-ears.

Make 5.

PIECING THE SETTING TRIANGLES

Handle all of the triangles carefully to avoid stretching the bias edges.

1. Sew a batik 5¼" triangle to one short side of a floral 4⅞" triangle. Press toward the batik triangle. Repeat on the other short side. Make a total of eight units.

Make 8.

2. Sew dark purple 5¼" triangles to opposite ends of the unit from step 1 as shown. Press the seam allowances toward the dark purple triangles. Make eight units.

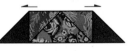

Make 8.

3. Sew the long edge of a dark purple 4⅞" triangle to the short side of a unit from step 2. Press the seam allowances toward the dark purple triangle. Make eight setting-triangle units. Trim the dog-ears.

Make 8.

4. Sew a floral 5¼" triangle to a batik 5¼" triangle along the long edge. Press the seam allowances toward the floral triangle. Make four units. Trim the dog-ears.

Make 4.

5. Sew a dark purple 5¼" triangle to each batik edge of the unit from step 1. Press the seam allowances toward the dark purple triangles. Make four corner setting-triangle units. Trim the dog-ears.

Make 4.

ASSEMBLING THE QUILT TOP

1. Referring to the quilt assembly diagram, lay out the blocks and the setting and corner triangles into diagonal rows. Sew the pieces in each row together. Press the seam allowances open to reduce bulk. Sew the rows together, pressing the seam allowances open.

2. Referring to "Borders" (page 74), apply the batik print inner border using mitered corners. Or see "Multiple Mitered Borders" (page 76) to join the inner border and outer border before adding them to the quilt.

3. Add the outer borders using the border fabric and mitered corners.

QUILTING AND BINDING

1. Refer to "Preparing the Quilt Sandwich" (page 76) to layer the quilt top, batting, and backing.
2. Hand or machine quilt as desired.
3. Referring to "Binding" (page 77), use the remaining 2"-wide batik print strips to bind the edges of your quilt.

QUILTING SUGGESTION

I so loved the idea of the butterflies on this quilt that I didn't want to do anything that would detract from them. So, I kept the quilting simple by outline quilting the beige, dark purple, and floral pieces; stitching in the ditch around the border; and outline quilting the butterflies and flowers in the border. I think it would be fun to quilt butterflies in the patchwork and have even more of them fluttering across the quilt.

A TISKET, A TASKET, THREE-DIMENSIONAL BASKETS

This quilt would work wonderfully in a foyer or hall, where you have a narrow wall to decorate, or it can be used as a door hanging. Another great idea would be to make each block and create three separate little quilts using single framed blocks. Then you could hang the three of them together vertically or horizontally. What a fabulous way to spruce up a blank wall!

Finished quilt: 26½" x 64¾"
Finished block: 12½" x 12½"

MATERIALS

Yardage is based on 42"-wide fabric.

- 1⅓ yards of beige fabric for blocks and setting triangles
- 1⅛ yards of green print for blocks, border, and binding
- ⅞ yard of brown print for blocks, sashing, and inner border
- ¼ yard of dark green tone-on-tone print for blocks
- ¼ yard of burgundy tone-on-tone print for blocks
- ¼ yard of pink sateen for prairie points
- ¼ yard of green sateen for prairie points
- 2 yards of fabric for backing
- 34" x 72" piece of batting
- 3¼ yards of ¼"-wide Steam-A-Seam 2
- Placement guide (see page 9)

CUTTING

All measurements include ¼"-wide seam allowances.

From the beige fabric, cut:
- 2 strips, 3" x 42"; crosscut into:
  6 rectangles, 3" x 8"
  1 square, 3" x 3"
- 1 square, 19" x 19"; cut into quarters diagonally to yield 4 triangles
- 2 squares, 9¾" x 9¾"; cut each square in half diagonally to yield 4 triangles
- 1 square, 10⅞" x 10⅞"; cut in half diagonally to yield 2 triangles (1 is extra)
- 2 squares, 5⅞" x 5⅞"; cut each square in half diagonally to yield 4 triangles (1 is extra)
- 1 square, 6¼" x 6¼"; cut into quarters diagonally to yield 4 triangles (2 are extra)
- 1 strip, 3⅜" x 42"; crosscut into 5 squares, 3⅜" x 3⅜". Cut each square in half diagonally to yield 10 triangles (1 is extra).

From the green print, cut:
- 1 strip, 3⅜" x 42"; crosscut into 10 squares, 3⅜" x 3⅜". Cut each square in half diagonally to yield 20 triangles.

- 1 square, 5⅞" x 5⅞"; cut in half diagonally to yield 2 triangles (1 is extra)
- 2 squares, 1½" x 1½"
- 4 squares, 1⅞" x 1⅞"; cut each square in half diagonally to yield 8 triangles
- 5 strips, 3" x 42"
- 5 strips, 2" x 42"

From the brown print, cut:
- 1 strip, 3⅜" x 42"; crosscut into 8 squares, 3⅜" x 3⅜". Cut each square in half diagonally to yield 16 triangles.
- 8 strips, 1½" x 42"; crosscut 4 strips into 12 rectangles, 1½" x 13"
- 1 square, 8⅜" x 8⅜"; cut in half diagonally to yield 2 triangles (1 is extra)
- 1 square, 5⅞" x 5⅞"; cut in half diagonally to yield 2 triangles (1 is extra)
- 1 basket handle using the pattern (page 28)

From the dark green tone-on-tone print, cut:
- 3 squares, 3⅜" x 3⅜"; cut each square in half diagonally to yield 6 triangles (1 is extra)

From the burgundy tone-on-tone print, cut:
- 1 strip, 3⅜" x 42"; crosscut into 7 squares, 3⅜" x 3⅜". Cut each square in half diagonally to yield 14 triangles (1 is extra).

From the pink sateen, cut:
- 2 strips, 3¼" x 42"; crosscut into 19 squares, 3¼" x 3¼"

From the green sateen, cut:
- 1 square, 7" x 7"
- 5 squares, 3¼" x 3¼"

PIECING BLOCK A

1. Referring to "Making Prairie Points" (page 8), make five pink sateen prairie points from the 3¼" squares and one green sateen prairie point from the 7" square.
2. Referring to "Fusing Prairie Points" (page 9), fuse the five pink sateen prairie points to five green print 3⅜" triangles and the green sateen prairie point to a green print 5⅞" triangle.

Pieced by Karen Sievert; quilted by Vicki Maloney

3. Sew a burgundy 3⅜" triangle to a pink sateen prairie-point unit along the long edge. Press the seam allowances toward the burgundy triangle. Make five units. Trim the dog-ears.

Make 5.

4. Sew a brown 5⅞" triangle to the green sateen prairie-point unit along the long edge. Press the seam allowances toward the brown print. Trim the dog-ears.

Make 1.

5. Sew a green print 3⅜" triangle to each short side of a beige 6¼" triangle. Press the seam allowances toward the green print triangles. Make two units total. Trim the dog-ears.

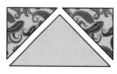

Make 2.

6. Sew a beige 3⅜" triangle to a green print 3⅜" triangle along the long edge. Press the seam allowances toward the green print triangle. Make two units total. Trim the dog-ears.

Make 2.

7. Sew two units from step 3 together as shown. Press the seam allowances open.

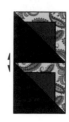

8. Lay out the large green prairie-point unit, the unit from step 7, and one unit from step 5 into a horizontal row. Sew the row together. Press the seam allowances open.

9. Lay out one unit from step 5, two units from step 6, three pink sateen prairie-point units, and the beige 3" square. Sew into rows, pressing the seam allowances open. Sew the rows together.

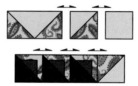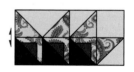

10. Sew the unit from step 9 to the top of the unit from step 8. Press the seam allowances open.

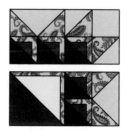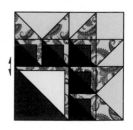

A TISKET, A TASKET, THREE-DIMENSIONAL BASKETS

11. Arrange two brown 3⅜" triangles and two beige 3" x 8" rectangles as shown. Sew each brown triangle to a rectangle, making two units. Press the seam allowances toward the brown print. Trim the dog-ears.

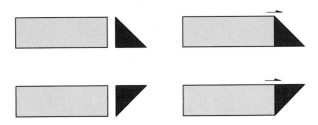

12. Sew the two units from step 11 to the sides of the unit from step 10. Press the seam allowances toward the rectangles.

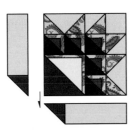 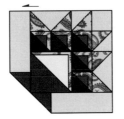

13. Sew a beige 5⅞" triangle to the bottom edge of the unit from step 12 to complete the block. Press the seam allowances toward the beige triangle. Trim the dog-ears.

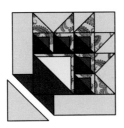 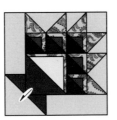

PIECING BLOCK B

1. Make six pink sateen prairie points from the 3¼" squares and fuse them to six green print 3⅜" triangles.
2. Sew a brown print 3⅜" triangle to a prairie-point unit from step 1 along the long edge. Press the seam allowances toward the brown print triangle. Make six units. Trim the dog-ears.

Make 6.

3. Arrange four brown print 3⅜" triangles and the six units from step 2 as shown. Sew into rows. Press the seam allowances open. Sew the rows together, again pressing the seam allowances open. Trim the dog-ears.

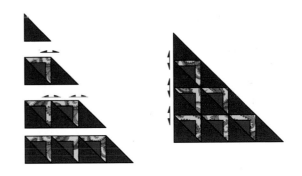

4. Fold the beige 10⅞" triangle in half to make a crease. Center the basket handle on the triangle and hand or machine appliqué it in place.

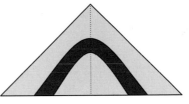

5. Sew the basket-handle unit to the long edge of the unit from step 3. Press the seam allowances toward the handle unit. Trim the dog-ears.

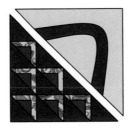 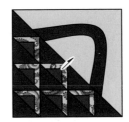

6. Refer to step 11 of "Piecing Block A" (page 25) to make one unit and one reversed unit using two brown 3⅜" triangles and two beige 3" x 8" rectangles.

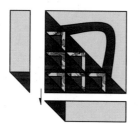

7. Sew the units from step 6 to the sides of the unit from step 5. Press the seam allowances toward the rectangles.

8. Sew a beige 5⅞" triangle to the bottom edge of the unit from step 7. Press the seam allowances toward the beige triangle. Trim the dog-ears.

PIECING BLOCK C

1. Make eight pink sateen and five green sateen prairie points from the 3¼" squares.
2. Fuse the following:
   • Five pink sateen prairie points to five beige 3⅜" triangles
   • Three pink sateen prairie points to three green print 3⅜" triangles
   • Two green sateen prairie points to two beige 3⅜" triangles
   • Three green sateen prairie points to three dark green tone-on-tone 3⅜" triangles

3. Sew a burgundy 3⅜" triangle to a pink/beige prairie-point unit along the long edge. Press the seam allowances toward the burgundy triangle. Make five units. Trim the dog-ears.

Make 5.

4. Sew a dark green tone-on-tone 3⅜" triangle to a green/beige prairie-point unit along the long edge. Press the seam allowances toward the dark green triangle. Make two. Sew a burgundy 3⅜" triangle to a pink/green print prairie-point unit along the long edge. Press the seam allowances toward the burgundy triangle. Make three. Trim the dog-ears.

Make 2.

Make 3.

5. Arrange the units from steps 3 and 4, and the three remaining prairie-point units as shown. Sew into rows, pressing the seam allowances open. Sew the rows together, again pressing the seam allowances open to reduce bulk.

A TISKET, A TASKET, THREE-DIMENSIONAL BASKETS

6. Sew a brown print 8⅜" triangle to the unit from step 5 along the long edge. Press the seam allowances toward the brown triangle. Trim the dog-ears.

7. Refer to step 11 of "Piecing Block A" (page 25) to make one unit and one reversed unit with two brown 3⅜" triangles and two beige 3" x 8" rectangles.

8. Sew the units from step 7 to the sides of the unit from step 6. Press the seam allowances toward the rectangles.

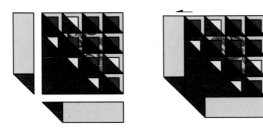

9. Sew a beige 5⅞" triangle to the bottom edge of the unit from step 8 to complete the block. Press the seam allowances toward the beige triangle. Trim the dog-ears.

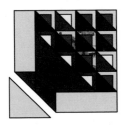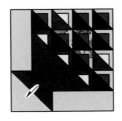

ASSEMBLING THE QUILT TOP

1. Refer to the quilt assembly diagram to lay out the three basket blocks, the brown 1½" x 13" rectangles, the green print 1½" squares and 1⅞" triangles, the beige 19" side setting triangles, and the beige 9¾" corner setting triangles into diagonal rows as shown. Sew the pieces in each row together. Press the seam allowances toward the brown print sashing. Sew the rows together. Press the seam allowances toward the sashing strips.

2. Join two brown 1½" x 42" strips together end to end to make one long strip. Make two.
3. Referring to "Borders" (page 74), apply the brown 1½"-wide border strips to the quilt top and bottom, and then to the sides.
4. Join two green print 3" x 42" strips together end to end to make one long strip. Make two long strips.
5. Apply the green print 3"-wide border strips to the quilt top and bottom, and then to the sides.

QUILTING AND BINDING

1. Refer to "Preparing the Quilt Sandwich" (page 76) to layer the quilt top, batting, and backing.
2. Hand or machine quilt as desired.
3. Referring to "Binding" (page 77), use the 2"-wide green print strips to bind the edges of your quilt.

QUILTING SUGGESTION

My hat's off to Vicki. She did a fabulous job of quilting this quilt. I love the way she filled the setting and corner triangles with her stunning feather work. And the use of a wool batting allows her quilting skills to stand out even more. Not only is wool batting great for warmth, but it's also wonderful for highlighting the quilting in a quilt. The fibers in the batting puff up in the lightly quilted areas, creating the look of trapunto. I love, love, love it!

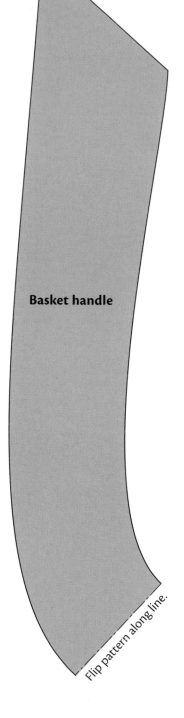

Basket handle

Flip pattern along line.

COLOR, COLOR, ALL AROUND

I've always been drawn to border-print fabrics. They can really create dynamic and dramatic quilts. Because this quilt would be equally beautiful with a non-border print, I've provided instructions for both options. The choice is yours! I think this is my favorite quilt in the book. Maybe it's the rich colors, or perhaps it's the way you can turn the blocks to come up with so many different designs. Whatever it is, this quilt sings to my soul. Have fun with this one, but watch out—it could be addictive.

Finished quilt: 50½" x 50½"
Finished block: 6" x 6"

MATERIALS

Yardage is based on 42"-wide fabric unless otherwise indicated.

▸ 5 yards of border print *OR* 3 yards of multicolored print for blocks, border, and binding*

▸ ½ yard of medium blue fabric for blocks

▸ ⅓ yard of dark blue fabric for blocks

▸ ⅓ yard *each* of medium gold, medium green, and medium red fabric for blocks and prairie points

▸ ¼ yard *each* of dark gold, dark green, and dark red fabric for blocks

▸ 3⅓ yards of fabric for backing

▸ 58" x 58" piece of batting

▸ Placement guide (see page 9)

▸ 11½ yards of ¼"-wide Steam-A-Seam 2

▸ Template plastic for fussy cutting border print

▸ Mirrors for fussy cutting**

*The border print should have at least four lengthwise repeats. If you use a non–border print, I suggest choosing a print with at least four colors.

**Magic Mirrors for Quilters and Quilter's Design Mirrors are available at quilt shops and online. They are very helpful for fussy cutting, but are not essential.

CUTTING

All measurements include ¼"-wide seam allowances.

From the medium blue fabric, cut:

• 2 strips, 2⅞" x 42"; crosscut into 18 squares, 2⅞" x 2⅞". Cut each square in half diagonally to yield 36 triangles.

• 3 strips, 2⅝" x 42"; crosscut into 36 squares, 2⅝" x 2⅝"

From *each* of the medium gold, medium green, and medium red fabrics, cut:

• 1 strip, 2⅞" x 42"; crosscut into 12 squares, 2⅞" x 2⅞". Cut each square in half diagonally to yield 24 triangles (72 total).

• 2 strips, 2⅝" x 42"; crosscut into 24 squares, 2⅝" x 2⅝" (72 total)

Continued on page 30

From the dark blue fabric, cut:

- 3 strips, 2⅞" x 42"; crosscut into 36 squares, 2⅞" x 2⅞". Cut each square in half diagonally to yield 72 triangles.

From *each* of the dark gold, dark green, and dark red fabrics, cut:

- 2 strips, 2⅞" x 42"; crosscut into 24 squares, 2⅞" x 2⅞". Cut each square in half diagonally to yield 48 triangles (144 total).

If Using a Border Print

From the border print, cut:

- 2 yards and set aside
- 36 triangles*
- 6 binding strips, 2" x 42"

See "Fussy Cutting Triangles" at right for detailed cutting instructions.

If Using a Multicolored Print

From the multicolored print, cut:

- 4 strips, 7½" x 70", on the *lengthwise* grain**
- 18 squares, 6⅞" x 6⅞"; cut each square in half diagonally to yield 36 triangles
- 6 binding strips, 2" x 42"

** *This allows for borders with mitered corners.*

PIECING THE BLOCKS

1. Referring to "Making Prairie Points" (page 8), make 36 medium blue, 24 medium gold, 24 medium green, and 24 medium red prairie points from the 2⅝" squares.
2. Referring to "Fusing Prairie Points" (page 9), fuse the following:
   - 36 medium blue prairie points to 36 dark blue 2⅞" triangles
   - 24 medium gold prairie points to 24 dark gold 2⅞" triangles
   - 24 medium green prairie points to 24 dark green 2⅞" triangles
   - 24 medium red prairie points to 24 dark red 2⅞" triangles

FUSSY CUTTING TRIANGLES

To get the same look as the quilt shown in the photograph, you'll need to fussy cut 36 triangles from the border print.

1. Cut a 6⅞" square from template plastic. Cut the square in half diagonally to yield two triangles.
2. Lay the border print flat, and use mirrors to determine an area that you like for the center square. The motif you choose will be repeated four times to create the kaleidoscope look at the center of the quilt.

3. Slide the triangle cut from template plastic into place between the mirrors and trace the motif onto the plastic. This will be your placement guide.

4. Remove the mirrors and trace around the triangle with a chalk pencil. Find the next identical motif, place the template on top, aligning the motifs, and again trace around the triangle. Repeat, tracing 36 triangles; cut on the marked lines.

Pieced and quilted by Karen Sievert

3. Sew a dark blue triangle to a blue prairie-point unit from step 2 along the long edges. Press the seam allowances toward the dark blue triangle. Make 36. Trim the dog-ears.

Make 36.

4. Lay out two units from step 3 and one medium blue 2⅞" triangle into a row. Sew the row together. Press the seam allowances open. Make 12 units.

Make 12.

5. Sew medium blue 2⅞" triangles to two sides of a blue prairie-point unit from step 3 as shown. Press the seam allowances toward the medium blue triangles. Make 12 units. Trim the dog-ears.

Make 12.

6. Sew a unit from step 4 to a unit from step 5. Press the seam allowances open. Make 12 units.

Make 12.

7. Sew a print 6⅞" triangle to the unit from step 6 along the long edge to finish the block. Press the seam allowances toward the unpieced triangle. Make 12 blocks. Trim the dog-ears.

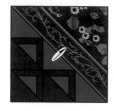

Make 12.

8. Repeat steps 3–7 to make eight gold, eight green, and eight red blocks.

ASSEMBLING THE QUILT TOP

Refer to the quilt assembly diagram to lay out the blocks in six rows of six blocks each, alternating colors as shown or in a manner that's pleasing to you. Sew the blocks in each row together. Press the seam allowances in opposite directions from row to row. Sew the rows together. Press the seam allowances in one direction.

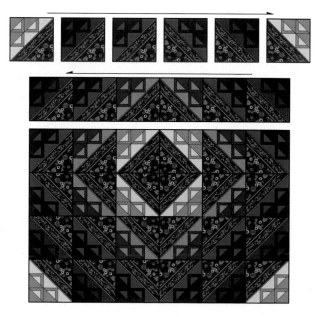

ADDING BORDERS

Follow the steps below for using a border print. For a non–border print, refer to "Borders" (page 74) to add the 7½"-wide border strips to the quilt top using either butted or mitered corners.

1. Choose the four stripes you like for the outer borders and cut four strips, 2 yards long. Be sure to add ¼" seam allowances on both sides of the stripe.
2. Experiment with the borders by laying them alongside the quilt top. You can move them from side to side, folding the corners at 45° angles, to assess where the most advantageous miter will be. By cutting the border strips 2 yards long, you have excess length to work with in making that determination.
3. Referring to "Borders with Mitered Corners" (page 75), add the border strips to the quilt top and press.

QUILTING AND BINDING

1. Refer to "Preparing the Quilt Sandwich" (page 76) to layer the quilt top, batting, and backing.
2. Hand or machine quilt as desired.
3. Referring to "Binding" (page 77), use the 2"-wide border-fabric strips to bind the edges of your quilt.

QUILTING SUGGESTION

How you lay out your blocks will be significant in deciding how to quilt this quilt. In some possible layouts, large squares will be formed that would be stunning with a motif filling the space and a matching pattern stitched in the border. In some, a simple outline stitch of the patchwork would work best. There are probably as many ways to quilt this little gem as there are layouts for it. My advice—just have fun with it.

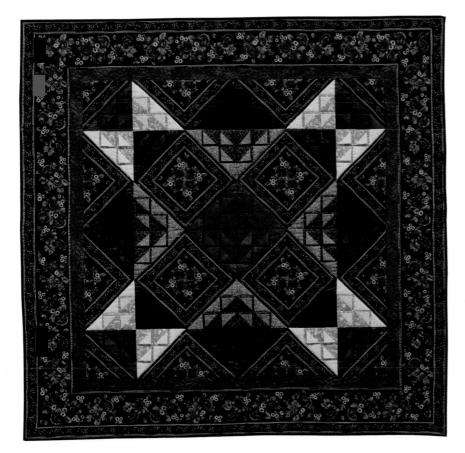

"Color, Color II," 50½" x 50½".
In this quilt, the block setting creates a kaleidoscope effect where the fussy-cut triangles meet. The prairie-point sections create yet another secondary design. For more layout ideas visit www.theniftyneedle.com. Click on "workshops," then "Color, Color" for more designs.

BEAR'S PAW—TABLE RUNNER

I've always loved the Bear's Paw block. And it seems to be tailor-made for prairie points! I couldn't stop sewing these blocks, so I designed two variations of this project—a table runner with 14" blocks and a lap quilt with 8¾" blocks. The lap quilt has a fantastic flying-geese border made with prairie points. You might think that making the border would be hard, but it's actually easier to make with prairie points than with regular piecing. Just wait and see!

Finished table runner: 22½" x 58½"
Finished block: 14" x 14"

MATERIALS

Yardage is based on 42"-wide fabric.

▸ 1½ yards of pink print for prairie points, border, and binding
▸ ⅝ yard of black solid for blocks and sashing strips
▸ ⅓ yard of floral print for blocks
▸ ¼ yard of dark pink print for blocks
▸ 2 yards of fabric for backing
▸ 30" x 66" piece of batting
▸ 5 yards of ¼"-wide Steam-A-Seam 2
▸ Placement guide (see page 9)

CUTTING

All measurements include ¼"-wide seam allowances.

From the pink print, cut:
• 4 strips, 2⅝" x 42"; crosscut into 48 squares, 2⅝" x 2⅝"
• 5 strips, 4½" x 42"
• 5 strips, 2" x 42"

From the dark pink print, cut:
• 2 strips, 2⅞" x 42"; crosscut into 24 squares, 2⅞" x 2⅞". Cut each square in half diagonally to yield 48 triangles.

From the black solid, cut:
• 2 strips, 2⅞" x 42"; crosscut into 24 squares, 2⅞" x 2⅞". Cut each square in half diagonally to yield 48 triangles.
• 5 strips, 2½" x 42"; crosscut into:
    4 rectangles, 2½" x 14½"
    12 rectangles, 2½" x 6½"
    12 squares, 2½" x 2½"

From the floral print, cut:
• 2 strips, 4½" x 42"; crosscut into:
    12 squares, 4½" x 4½"
    3 squares, 2½" x 2½"

PIECING THE BLOCKS

1. Referring to "Making Prairie Points" (page 8), make 48 prairie points from the pink print 2⅝" squares.

2. Referring to "Fusing Prairie Points" (page 9), fuse the 48 prairie points to the 48 dark pink print 2⅞" triangles.

3. Sew a black 2⅞" triangle to a prairie-point unit along the long edge. Press the seam allowances toward the black triangle. Make a total of 48. Trim the dog-ears.

Make 48.

4. Lay out two units from step 3 as shown and sew them together. Press the seam allowances toward the right side. Make a total of 12 units.

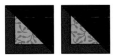

Make 12.

5. Lay out two units from step 3 and a black 2½" square as shown and sew together into a row. Press the seam allowances upward. Make a total of 12 units.

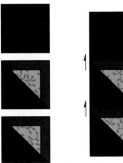

Make 12.

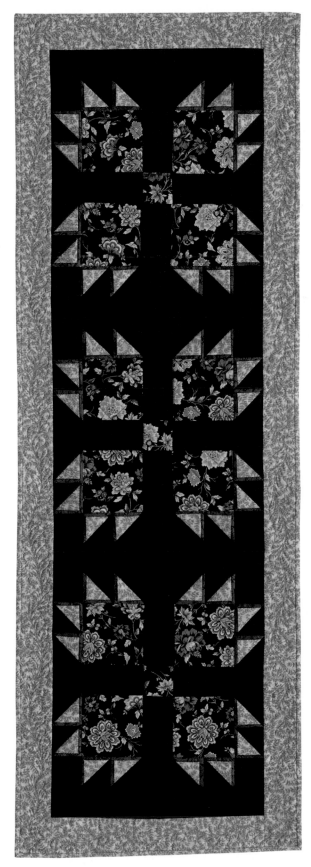

Pieced and quilted by Karen Sievert

6. Sew a unit from step 4 to the top of a floral print 4½" square. Press the seam allowances toward the floral square. Sew a unit from step 5 to the left side of the floral square. Press. Make a total of 12 units.

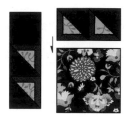

Make 12.

7. Lay out four units from step 6, one floral 2½" square, and four black 2½" x 6½" rectangles. Sew into rows. Press the seam allowances toward the black fabric. Sew the rows together to complete the block. Press. Make three blocks.

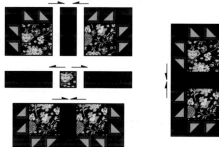

Make 3.

ASSEMBLING THE TABLE RUNNER

1. Refer to the quilt assembly diagram to lay out the blocks and the black 2½" x 14½" rectangles. Sew together. Press the seam allowances toward the black rectangles.

2. Sew three pink print 4½" x 42" strips together end to end to make one long strip for the side borders.
3. Referring to "Borders" (page 74), sew the pink 4½"-wide border strips to the top and bottom of the table runner. Then add the side borders.

QUILTING AND BINDING

1. Refer to "Preparing the Quilt Sandwich" (page 76) to layer the quilt top, batting, and backing.
2. Hand or machine quilt as desired.
3. Referring to "Binding" (page 77), use the 2"-wide pink strips to bind the edges of your quilt.

QUILTING SUGGESTION

Once again, I adhered to the "keep it simple" principle in quilting this table runner. A ¼" outline of the patchwork and a feather in the border meant I had this little beauty gracing the table in no time at all. Don't you just love quick, easy projects?

BEAR'S PAW—LAP QUILT

I love curling up with this quilt. I've combined the beloved Bear's Paw block with another favorite, Log Cabin. The pink and green colors really pop against the black background, and the prairie points add a wonderful extra element. The Flying-Geese border is super simple, super fun, and super effective!

Finished quilt: 46" x 65"

Finished Bear's Paw block: 8¾" x 8¾"

Finished Log Cabin block: 4¼" x 4¼"

MATERIALS

Yardage is based on 42"-wide fabric.

- ► 2⅔ yards of black solid for blocks, border, and binding
- ► 1⅛ yards of light pink print for prairie points and blocks
- ► 1 yard of floral print for blocks and setting triangles
- ► ⅞ yard of medium pink print for blocks
- ► ¼ yard of light green print for blocks
- ► ¼ yard of green solid for blocks
- ► ¼ yard of dark pink print for blocks
- ► 3 yards of fabric for backing
- ► 54" x 72" piece of batting
- ► 32 yards of ¼"-wide Steam-A-Seam 2
- ► Placement guide (see page 9)

CUTTING

All measurements include ¼"-wide seam allowances.

From the light pink print, cut:

- 6 strips, 1⅝" x 42"; crosscut into 128 squares, 1⅝" x 1⅝"
- 11 strips, 2" x 42"; crosscut into 208 squares, 2" x 2"
- 3 strips, 1" x 42"

From the medium pink print, cut:

- 4 strips, 2⅛" x 42"; crosscut into 64 squares, 2⅛" x 2⅛". Cut each square in half diagonally to yield 128 triangles.
- 7 strips, 2" x 42"; crosscut into:
    208 rectangles, 2" x 1¼"
    4 squares, 2" x 2"
- 4 strips, 1" x 42"

From the black solid, cut:

- 7 strips, 1¾" x 42"; crosscut into:
    54 squares, 1¾" x 1¾"
    32 rectangles, 1¾" x 4¼"
- 4 strips, 2⅛" x 42"; crosscut into 64 squares, 2⅛" x 2⅛". Cut each square in half diagonally to yield 128 triangles.
- 3 strips, 1" x 42"; crosscut into 4 rectangles, 1" x 9¼", and 3 rectangles, 1" x 18½"
- 1 strip, 7⅜" x 42"; crosscut into 5 squares, 7⅜" x 7⅜". Cut each square into quarters diagonally to yield 20 triangles (2 are extra).

Continued on page 38

- 2 squares, 4" x 4"; cut each square in half diagonally to yield 4 triangles
- 1 strip, 1⅝" x 42"; crosscut into 22 squares, 1⅝" x 1⅝"
- 6 strips, 6½" x 42"
- 6 strips, 2" x 42"

From the floral print, cut:
- 1 strip, 1¾" x 42"; crosscut into 8 squares, 1¾" x 1¾"
- 3 strips, 3" x 42"; crosscut into 32 squares, 3" x 3"
- 6 strips, 1" x 42"
- 1 strip, 7⅜" x 42"; crosscut into 5 squares, 7⅜" x 7⅜". Cut each square into quarters diagonally to yield 20 triangles (2 are extra).
- 1 strip, 4" x 42"; crosscut into 6 squares, 4" x 4". Cut each square in half diagonally to yield 12 triangles.

From the light green print, cut:
- 4 strips, 1" x 42"

From the green solid, cut:
- 5 strips, 1" x 42"

From the dark pink print, cut:
- 5 strips, 1" x 42"

MAKING THE BEAR'S PAW BLOCKS

1. Referring to "Making Prairie Points" (page 8), make 128 prairie points from the light pink 1⅝" squares.
2. Referring to "Fusing Prairie Points" (page 9), fuse the 128 light pink prairie points to the 128 medium pink 2⅛" triangles.
3. Sew a black 2⅛" triangle to a prairie-point unit along the long edge. Press the seam allowances toward the black triangle. Make 128. Trim the dog-ears.

Make 128.

4. Lay out two units from step 3 as shown and sew them together. Press toward the right side. Make 32 units.

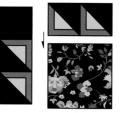

Make 32.

5. Lay out two units from step 3 and a black 1¾" square as shown. Sew together into a row. Press the seam allowances toward the top of the unit. Make 32 units.

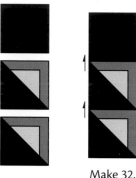

Make 32.

6. Sew a unit from step 4 to the top of a floral 3" square. Press the seam allowances toward the floral square. Sew a unit from step 5 to the left side of the unit. Press toward the floral square. Make 32 units.

Make 32.

BEAR'S PAW—LAP QUILT

7. Lay out four units from step 6, one floral 1¾" square, and four black 1¾" x 4¼" rectangles. Sew into rows. Press the seam allowances toward the black fabric. Sew the rows together to complete the block. Press the seam allowances toward the black fabric. Make eight blocks.

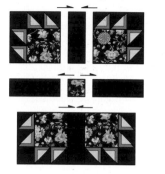

Make 8.

MAKING THE LOG CABIN A BLOCKS

The A blocks will be placed along the top, bottom, and sides of the quilt center. Press all seam allowances toward the strip just added.

1. Place one light pink 1" strip on a black 1¾" square with right sides together and raw edges aligned. Stitch together as shown. Trim the strip even with the square. Press.

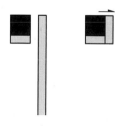

2. Rotate the unit just made a quarter turn to the left and repeat step 1.

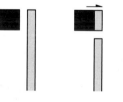

3. Repeat the process to add two light green strips.

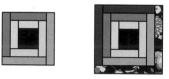

4. Continue rotating and adding two strips in the following order: medium pink, green solid, dark pink, and floral print.

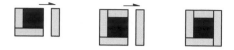

5. Repeat steps 1–4 to make a total of 18 Log Cabin A blocks.

MAKING THE LOG CABIN B BLOCKS

The B blocks will be placed in the corners. They are made in the same way as the Log Cabin A blocks, by sewing the strip and trimming. However, the order of sewing is different. Press all the seam allowances toward the strip just added.

1. Sew a light pink 1" strip to a black 1¾" square. Press and trim even with the square.

2. Use the light green strip and sew to the opposite side. Press. Sew a light green strip to the two remaining sides.

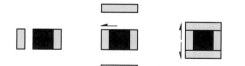

3. Sew the medium pink strip next to the light pink strip. Sew the green solid strip on the opposite end, and then on the two remaining sides. Continue in this manner, substituting dark pink, and then the floral print to complete the block.

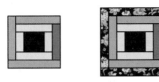

4. Repeat steps 1–3 to make a total of four Log Cabin B blocks.

ASSEMBLING THE QUILT TOP

1. Sew a Bear's Paw block to each side of a black 1" x 9¼" rectangle to make a row. Press the seam allowances toward the black rectangle. Make four rows.

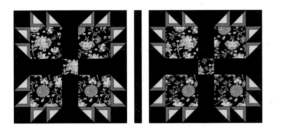

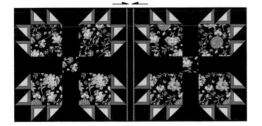

2. Lay out the rows from step 1 and the three black 1" x 18½" rectangles. Sew together, pressing the seam allowances toward the black fabric.

Assembling the Log Cabin Border

1. Lay out one Log Cabin B block, two floral 4" triangles, and one floral 7⅜" triangle as shown. Sew together, pressing the seam allowances toward the floral triangles. Trim the dog-ears. Make two units.

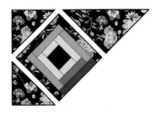 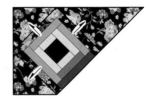

Make 2.

2. Lay out one black 7⅜" triangle, one Log Cabin A block, and one floral 7⅜" triangle. Sew together. Press the seam allowances toward the triangles. Make six units.

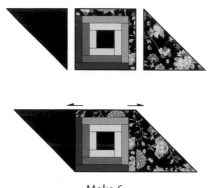

Make 6.

3. Lay out one Log Cabin B block, two floral 4" triangles, and one black 7⅜" triangle as shown. Sew together. Press the seam allowances toward the triangles. Make two units.

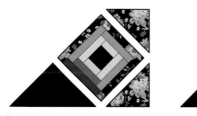

Make 2.

4. Lay out one unit from step 1, three units from step 2, and one unit from step 3. Sew together. Press the seam allowances in either direction. Make two units for the top and bottom borders. Trim and straighten the edges as needed, making sure to leave ¼" beyond the points of the blocks.

Make 2.

5. Lay out one Log Cabin A block, one floral 4" triangle, one floral 7⅜" triangle, and one black 4" triangle as shown. Sew together. Press the seam allowances toward the triangles. Trim the dog-ears. Make two units.

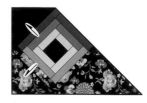

Make 2.

6. Lay out one Log Cabin A block, one floral 7⅜" triangle, and one black 7⅜" triangle as shown. Sew together. Press the seam allowances toward the triangles. Make eight units.

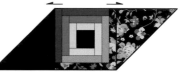

Make 8.

7. Lay out a Log Cabin A block, a floral 4" triangle, a black 4" triangle, and a black 7⅜" triangle as shown. Sew together. Press the seam allowances toward the triangles. Make two units.

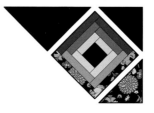

Make 2.

8. Lay out one unit from step 5, four units from step 6, and one unit from step 7. Sew together. Press the seam allowances in either direction. Make two units for the side borders. Trim and straighten the edges as needed, making sure to leave ¼" beyond the points of the blocks.

Make 2.

9. Sew the pieced side borders to each long side of the quilt top. The black triangles should be on the inside of the quilt and the floral triangles along the outside. Sew the top and bottom borders to the quilt, having the black triangles on the inside and the floral triangles along the outside.

Piecing the Flying-Geese Border

1. Make 208 prairie points from the light pink 2" squares.
2. Place a 1½"-long piece of the ¼" Steam-A-Seam 2 along the bottom long edge of a medium pink 1¼" x 2" rectangle; then fuse a prairie point to the rectangle.

3. Sew a mediuim pink rectangle to the unit from step 2 along the bottom edge. Place a piece of the ¼"-wide Steam-A-Seam 2 on the rectangle just sewn, line up a prairie point with the one above it, and fuse into place. Then sew another rectangle to the bottom edge. Continue in this manner to make two units with 40 prairic points each and two units with 64 prairie points each.

4. Sew a 2" medium pink square to each end of the border strips with 40 prairie points. Press the seam allowances toward the squares.
5. Sew the borders with 64 prairie points to the sides of the quilt top, and then sew the borders with 40 prairie points to the top and bottom. Press the seam allowances toward the Log Cabin border.

Adding the Outer Border

Referring to "Borders" (page 74), piece and add the black 6½"-wide border strips to the sides of the quilt top, and then to the top and bottom.

QUILTING AND BINDING

1. Refer to "Preparing the Quilt Sandwich" (page 76) to layer the quilt top, batting, and backing.
2. Hand or machine quilt as desired.
3. Referring to "Binding" (page 77), use the 2"-wide black strips to bind the edges of your quilt.

QUILTING SUGGESTION

I kept the quilting on this quilt fairly straightforward. I outline quilted the patchwork of the Bear's Paw blocks and the Log Cabin blocks. In the setting triangles I quilted miniature feathers. I pulled it all together by quilting more feathers in the outer border. This is one quilt I'll really enjoy snuggling under during those cold winter months!

SPURRED ON

When my son Travis was about five years old, he gave me a concert on Mother's Day. He got all dressed up in his jeans, cowboy boots, plaid shirt, and hat. He strapped on a guitar, put a George Strait disk in the CD player, and proceeded to sing *every single song* to me! I vowed then that one day I'd make him a cowboy quilt. Well, that day finally arrived. He's 15 now, but that's OK. Every time I see the quilt, it puts a smile in my heart. Of course, if you don't have a cowboy in your life, the design would be lovely with a large-scale floral or other theme print.

Finished quilt: 64" x 64"
Finished block: 10" x 10"

MATERIALS

Yardage is based on 42"-wide fabric.

- ▶ 1⅞ yards of red cowboy print for blocks, outer border, and binding
- ▶ 1⅓ yards of red tone-on-tone print for blocks
- ▶ 1 yard of beige star print for blocks and inner border
- ▶ 1 yard of beige tone-on-tone print for setting triangles
- ▶ 1 yard of red plaid for prairie points
- ▶ ⅞ yard of large-scale cowboy print for center square
- ▶ ¼ yard of black cowboy print for blocks
- ▶ 4 yards of fabric for backing
- ▶ 72" x 72" piece of batting
- ▶ 17 yards of ¼"-wide Steam-A-Seam 2
- ▶ Placement guide (see page 9)

CUTTING

All measurements include ¼"-wide seam allowances.

From the red plaid, cut:
- 11 strips, 2⅝" x 42"; crosscut into 160 squares, 2⅝" x 2⅝"

From the red tone-on-tone print, cut:
- 6 strips, 2⅞" x 42"; crosscut into 80 squares, 2⅞" x 2⅞". Cut each square in half diagonally to yield 160 triangles.
- 2 strips, 10⅞" x 42"; crosscut into 6 squares, 10⅞" x 10⅞". Cut each square in half diagonally to yield 12 triangles.

From the beige star print, cut:
- 5 strips, 2⅞" x 42"; crosscut into 60 squares, 2⅞" x 2⅞". Cut each square in half diagonally to yield 120 triangles.
- 2 strips, 2½" x 42"; crosscut into 20 squares, 2½" x 2½"
- 4 strips, 2¾" x 42"; crosscut into:
     2 strips, 2¾" x 24¼"
     2 strips, 2¾" x 28¾"

Continued on page 46

Pieced and quilted by Karen Sievert

From the red cowboy print, cut:
- 2 strips, 6⅞" x 42"; crosscut into 6 squares, 6⅞" x 6⅞". Cut each square in half diagonally to yield 12 triangles.
- 8 strips, 4" x 42"
- 7 strips, 2" x 42"

From the black cowboy print, cut:
- 1 strip, 6⅞" x 42"; crosscut into 4 squares, 6⅞" x 6⅞". Cut each square in half diagonally to yield 8 triangles.

From the large-scale cowboy print for center square, cut:
- 1 square, 24¼" x 24¼"

From the beige tone-on-tone print, cut:
- 2 squares, 15½" x 15½"; cut each square into quarters diagonally to yield 8 triangles
- 2 squares, 15⅛" x 15⅛"; cut each square in half diagonally to yield 4 triangles

PIECING THE BLOCKS

1. Referring to "Making Prairie Points" (page 8), make 160 prairie points from the plaid 2⅝" squares.
2. Referring to "Fusing Prairie Points" (page 9), fuse the 160 plaid prairie points to the 160 red tone-on-tone 2⅞" triangles.
3. Sew a beige star print 2⅞" triangle to a prairie-point unit along the long edge. Press the seam allowances toward the beige triangle. Make a total of 120 units.

Make 120.

4. Lay out one beige print 2½" square, three prairie-point units from step 3, and one prairie point from step 2. Sew together into a row. Press the seam allowances toward the beige print. Make a total of 20 units.

Make 20.

5. Lay out three prairie-point units from step 3 and one prairie point from step 2. Sew into a row. Press the seam allowances toward the beige print. Make a total of 20 units.

Make 20.

6. Sew a unit from step 5 to the short left side of a red cowboy print 6⅞" triangle and sew a unit from step 4 to the right side. Press the seam allowances toward the cowboy print. Make a total of 12 units.

 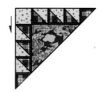

Make 12.

7. Sew a red tone-on-tone 10⅞" triangle to the long edge of a red cowboy unit from step 6. Press the seam allowances toward the red triangle. Make a total of 12 red Cowboy blocks.

Make 12.

8. Repeat step 6 with the black cowboy print triangles to make eight black cowboy units.

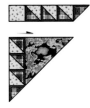

Make 8.

ASSEMBLING THE QUILT TOP

1. Sew the beige star print 2¾" x 24¼" strips to the top and bottom of the square center panel. Press the seam allowances toward the border strips. Sew the 2¾" x 28¾" strips to the sides. Press.

2. Lay out two black cowboy units and one red Cowboy block as shown. Sew the black units to each side of the Cowboy block. Press the seam allowances toward the red triangle. Make four units.

Make 4.

3. Referring to the quilt assembly diagram, sew a unit from step 2 to each side of the bordered center panel. Then sew a unit to the top and bottom of the panel. Press the seam allowances toward the bordered panel.

4. Lay out two red Cowboy blocks, two beige 15½" triangles, and one beige 15⅛" triangle. Sew together. Press the seam allowances toward the beige triangles. Make four units.

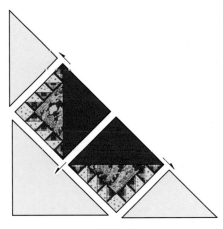

Make 4.

5. Sew units from step 4 to opposite sides of the unit from step 3, and then sew a unit to each of the remaining sides. Press seam allowances outward.

6. Sew two red cowboy print 4" x 42" strips together end to end to make one long strip. Make four strips.
7. Referring to "Borders" (page 74), sew the strips to the quilt top using butted corners.

QUILTING AND BINDING

1. Refer to "Preparing the Quilt Sandwich" (page 76) to layer the quilt top, batting, and backing.
2. Hand or machine quilt as desired.
3. Referring to "Binding" (page 77), use the 2"-wide red cowboy print strips to bind the edges of your quilt.

QUILTING SUGGESTION

No feathers or flowers for this cowboy! To stay with the masculine theme, I outline quilted the blocks, and then quilted parallel straight lines in the center panel and beige setting triangles. I added some curves by quilting a pumpking-seed design in the large red triangles. It's a lot of straight lines, but all in all, suitable for a young man.

POP STARS

When I was making "Star-Studded Sampler" (page 59), I fell in love with this block, called Rhode Island. The star within a star was so appealing that I decided to make a quilt using just that block. I wish I could have found a fabric that had a celebrity on it. That would really make this quilt a "pop star!" I think it's beautiful in green and pink, but would also be fabulous as a Christmas quilt in red and green. So many possibilities.

Finished quilt: 50½" x 50½"
Finished block: 12" x 12"

MATERIALS

Yardage is based on 42"-wide fabric.

- ► 1⅞ yards of dark floral print for blocks, outer border, and binding
- ► ⅞ yard of white print 1 for blocks (outer star)
- ► ¾ yard of green print for prairie points and inner border
- ► ½ yard of pink tone-on-tone print for prairie points
- ► ½ yard of dark green tone-on-tone print for blocks
- ► ½ yard of dark pink tone-on-tone print for blocks
- ► ⅜ yard of white print 2 for blocks (inner star)
- ► ⅓ yard of light floral print for blocks
- ► 3¼ yards of fabric for backing
- ► 58" x 58" piece of batting
- ► 10½ yards of ¼"-wide Steam-A-Seam 2
- ► Placement guide (see page 9)

CUTTING

All measurements include ¼"-wide seam allowances.

From the green print, cut:
- 4 strips, 3¾" x 42"; crosscut into 40 squares, 3¾" x 3¾"
- 4 strips, 2" x 42"

From the pink tone-on-tone print, cut:
- 4 strips, 3¾" x 42"; crosscut into 32 squares, 3¾" x 3¾"

From the dark green tone-on-tone print, cut:
- 2 strips, 5¼" x 42"; crosscut into 10 squares, 5¼" x 5¼". Cut each square into quarters diagonally to yield 40 triangles.
- 1 strip, 3¼" x 42"; crosscut into 10 squares, 3¼" x 3¼". Cut each square into quarters diagonally to yield 40 triangles.

From the dark pink tone-on-tone print, cut:
- 2 strips, 5¼" x 42"; crosscut into 8 squares, 5¼" x 5¼". Cut each square into quarters diagonally to yield 32 triangles.
- 1 strip, 3¼" x 42"; crosscut into 8 squares, 3¼" x 3¼". Cut each square into quarters diagonally to yield 32 triangles.

From the white print 2, cut:
- 2 strips, 2⅞" x 42"; crosscut into 18 squares, 2⅞" x 2⅞". Cut each square in half diagonally to yield 36 triangles.
- 2 strips, 1⅞" x 42"; crosscut into 36 squares, 1⅞" x 1⅞"

Continued on page 50

From the dark floral print, cut:
- 2 strips, 4⅞" x 42"; crosscut into 10 squares, 4⅞" x 4⅞". Cut each square in half diagonally to yield 20 triangles.
- 1 strip, 3⅜" x 42"; crosscut into 5 squares, 3⅜" x 3⅜"
- 5 strips, 6" x 42"
- 6 strips, 2" x 42"

From the white print 1, cut:
- 3 strips, 4⅞" x 42"; crosscut into 18 squares, 4⅞" x 4⅞". Cut each square in half diagonally to yield 36 triangles.
- 2 strips, 5¼" x 42"; crosscut into 9 squares, 5¼" x 5¼". Cut each square into quarters diagonally to yield 36 triangles.

From the light floral print, cut:
- 1 strip, 4⅞" x 42"; crosscut into 8 squares, 4⅞" x 4⅞". Cut each square in half diagonally to yield 16 triangles.
- 1 strip, 3⅜" x 42"; crosscut into 4 squares, 3⅜" x 3⅜"

PIECING THE BLOCKS

1. Referring to "Making Prairie Points" (page 8), make 40 prairie points from the green print 3¾" squares and 32 prairie points from the pink tone-on-tone print 3¾" squares.
2. Referring to "Fusing Prairie Points" (page 9), fuse the 40 green prairie points to the 40 dark green 5¼" triangles. Fuse the 32 pink prairie points to the 32 dark pink 5¼" triangles.
3. Sew a dark green 3¼" triangle to each short side of a white print 2 triangle. Press the seam allowances toward the dark green triangles. Make 20 units.

Make 20.

4. Lay out four units from step 3, four white print 2 squares, and one dark floral 3⅜" square into three horizontal rows as shown. Sew each row together. Press the seam allowances in the direction indicated. Sew the rows together. Press the seam allowances toward the center. Make five units.

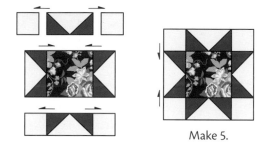

Make 5.

5. Sew white print 1 and dark floral 4⅞" triangles together along the long edge. Press the seam allowances toward the dark floral triangle. *Don't* trim the dog-ears; they give placement guides for the next step. Make 20 units.

Make 20.

6. Align the points and sides of a green prairie-point unit with one white side of a unit from step 5. Press the seam allowances toward the white triangle. Repeat on the adjacent side. Make 20 units.

Make 20.

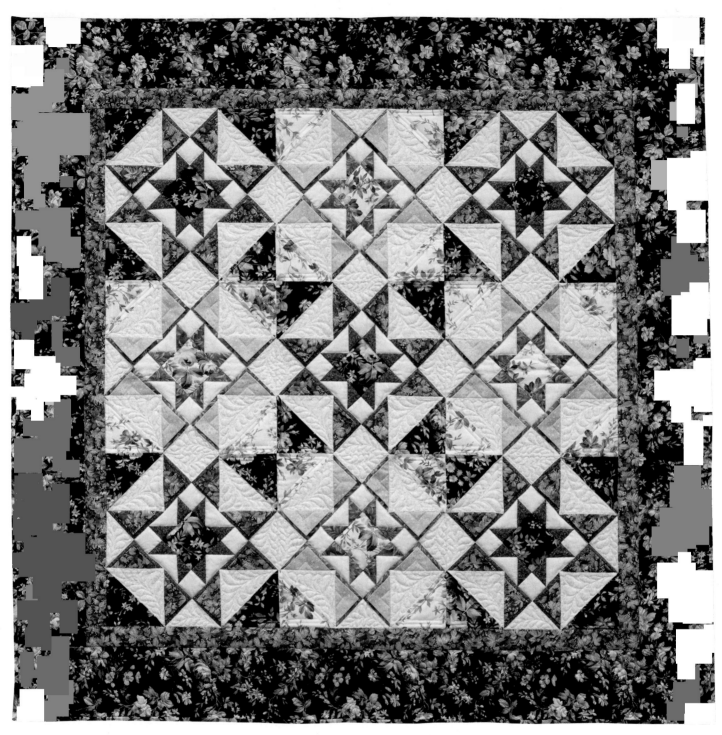

Pieced and quilted by Karen Sievert

7. Sew 5¼" white print 1 triangles to opposite sides of a unit from step 6. Press the seam allowances toward the white triangles. Make 10 units.

 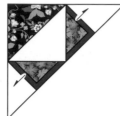

Make 10.

8. Lay out two units from step 6, two units from step 7, and one unit from step 4. Sew the two units from step 6 to opposite sides of the unit from step 4. Press the seam allowances open. Sew the two units from step 7 to the remaining two sides to complete the block. Press the seam allowances open. Make five blocks.

9. Repeat steps 3–8 to make four pink blocks, substituting dark pink for dark green, light floral for dark floral, and pink prairie-point units for green.

Make 4.

ASSEMBLING THE QUILT TOP

1. Refer to the quilt assembly diagram to lay out the blocks in three rows of three blocks each, alternating colors from row to row. Sew the blocks in each row together. Press the seam allowances in opposite directions from row to row. Sew the rows together. Press the seam allowances in one direction.

2. Referring to "Borders" (page 74), apply the green print 2"-wide border strips to the quilt top. Then apply the 6"-wide dark floral border strips to the quilt top, piecing three strips together first for the side borders.

QUILTING AND BINDING

1. Refer to "Preparing the Quilt Sandwich" (page 76) to layer the quilt top, batting, and backing.
2. Hand or machine quilt as desired.
3. Referring to "Binding" (page 77), use the 2"-wide dark floral strips to bind the edges of your quilt.

QUILTING SUGGESTION

I had a lot of fun quilting this one. I stitched in the ditch of the interior stars, and then stitched miniature feathers in the remaining white spaces. I repeated the feathers in a larger size in the border. The wool batting really highlights the feathers.

BLOOMING TREE OF LIFE

I couldn't wait to make a Tree of Life quilt—I just knew it would be fantastic with prairie points. They give the tree that little bit of added dimension so that it appears to be blooming. There's also all that wonderful space for quilting. A definite keeper, this quilt will be gracing my home for years and years to come.

Finished quilt: 72½" x 72½"
Finished block: 20" x 20"

MATERIALS

Yardage is based on 42"-wide fabric.

- 2⅞ yards of beige print for blocks and setting triangles
- 2½ yards of floral border print for tree trunks and inner border
- 1⅔ yards of light green print for blocks, outer border, and binding
- ⅝ yard of light pink print for prairie points and blocks
- ½ yard of dark green tone-on-tone print for blocks
- ½ yard of medium green print for blocks
- ½ yard of medium pink print for blocks
- ⅜ yard of dark pink tone-on-tone print for prairie points
- ⅜ yard of burgundy tone-on-tone print for prairie points
- 4½ yards of fabric for backing
- 80" x 80" piece of batting
- 14 yards of ¼"-wide Steam-A-Seam 2
- 6" square ruler
- Placement guide (see page 9)

CUTTING

All measurements include ¼"-wide seam allowances.

From the beige print, cut:
- 1 strip, 6⅞" x 42"; crosscut into 6 squares, 6⅞" x 6⅞". Cut each square in half diagonally to yield 12 triangles.
- 1 square, 20½" x 20½"
- 1 square, 29⅝" x 29⅝"; cut into quarters diagonally to yield 4 triangles
- 2 squares, 15⅛" x 15⅛"; cut each square in half diagonally to yield 4 triangles
- 2 strips, 8½" x 42"; crosscut into 8 pieces, 8½" x 8⅝"

From the medium pink print, cut:
- 3 strips, 2⅞" x 42"; crosscut into 36 squares, 2⅞" x 2⅞". Cut each square in half diagonally to yield 72 triangles.
- 1 strip, 2½" x 12"; crosscut into 4 squares, 2½" x 2½"

From the dark green tone-on-tone print, cut:
- 2 squares, 9¼" x 9¼"; cut each square into quarters diagonally to yield 8 triangles
- 2 strips, 2⅞" x 42"; crosscut into 24 squares, 2⅞" x 2⅞". Cut each square in half diagonally to yield 48 triangles.

From the medium green print, cut:
- 4 strips, 2⅞" x 42"; crosscut into 48 squares, 2⅞" x 2⅞". Cut each square in half diagonally to yield 96 triangles.
- 1 strip, 2½" x 12"; crosscut into 4 squares, 2½" x 2½"

From the light green print, cut:
- 4 strips, 2⅞" x 42"; crosscut into 48 squares, 2⅞" x 2⅞". Cut each square in half diagonally to yield 96 triangles.
- 4 squares, 2½" x 2½"
- 8 strips, 3" x 42"
- 8 strips, 2" x 42"

From the light pink print, cut:
- 3 strips, 2⅝" x 42"; crosscut into 48 squares, 2⅝" x 2⅝"
- 2 squares, 5¼" x 5¼"; cut each square into quarters diagonally to yield 8 triangles

From the dark pink tone-on-tone print, cut:
- 3 strips, 2⅝" x 42"; crosscut into 48 squares, 2⅝" x 2⅝"

From the burgundy tone-on-tone print, cut:
- 3 strips, 2⅝" x 42"; crosscut into 48 squares, 2⅝" x 2⅝"

From the floral border print, cut:
- 4 strips, 5¾" x 90", on the *lengthwise* grain*
- 4 *fussy-cut* rectangles, 3⅜" x 16⅛"**

*The width of your border strips may vary, depending on your fabric.

**These are for the tree trunks. Fussy cut a motif that you like.

PIECING THE BLOCKS

1. Referring to "Making Prairie Points" (page 8), make 48 prairie points *each* from the light pink, dark pink, and burgundy 2⅝" squares.
2. Referring to "Fusing Prairie Points" (page 9), fuse the prairie points as follows:
   - 48 light pink prairie points to 48 medium pink print 2⅞" triangles
   - 48 dark pink prairie points to 48 light green 2⅞" triangles
   - 48 burgundy prairie points to 48 medium green 2⅞" triangles
3. Fold a beige 8½" x 8⅝" piece in half to find the center vertically as shown. Using chalk, mark the center at both ends. Lay a 6" square ruler diagonally on the piece of fabric with the 45° line running through both center marks and the tip of the ruler aligned with the top edge. Rotary cut the triangles away at the top. Repeat for all eight pieces.

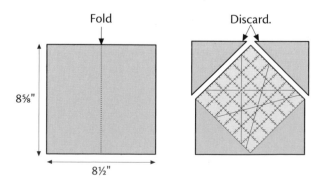

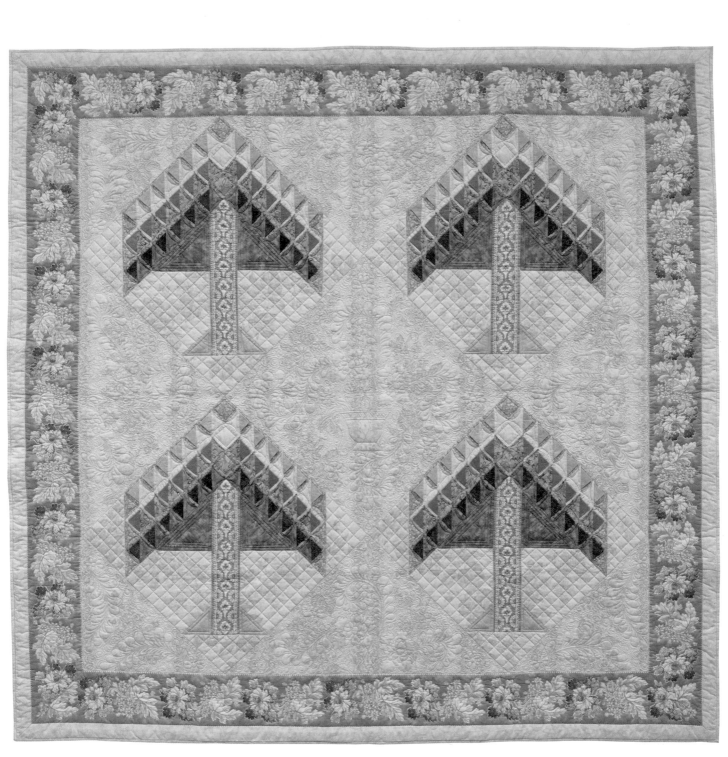

Pieced and quilted by Karen Sievert

BLOOMING TREE OF LIFE

4. Sew a dark green 9¼" triangle to an angled side of a beige house-shaped piece as shown. Make four and four reversed. Press the seam allowances toward the green triangle.

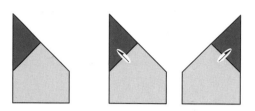

Make 4 of each.

5. Sew a light pink 5¼" triangle to the opposite side of each unit from step 4. Press the seam allowances toward the pink triangle.

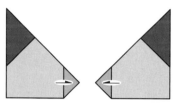

Make 4 of each.

6. Fold each border print 3⅜" x 16⅛" rectangle in half to find the center vertically and make a crease. Using a chalk pencil, mark the center at the top. Lay a 6" square ruler diagonally on the fabric with the 45° line running through the center mark and the crease; the tip of the ruler should align with the top edge. Rotary cut the triangles away at the top. Repeat for all four rectangles.

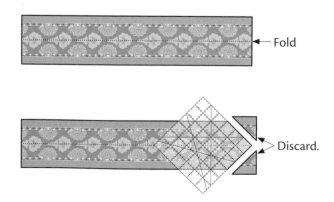

← Fold

Discard.

7. Sew a unit and a reversed unit from step 5 to each side of the border print 16⅛" piece as shown. Press away from the center piece. Make four units.

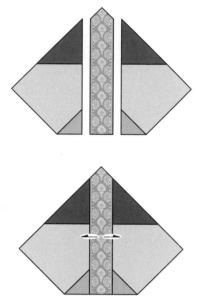

Make 4.

8. Sew a beige 6⅞" triangle to the bottom of each unit from step 7. Press toward the triangle.

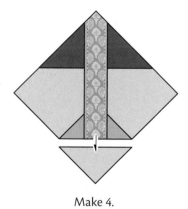

Make 4.

9. Sew a light green 2⅞" triangle to a light pink prairie-point unit along the long edge. Press the seam allowances toward the light green. Make 48 light green units. Repeat to make 48 units with medium green triangles sewn to

the dark pink prairie-point units and 48 units with the dark green triangles sewn to the burgundy prairie-point units.

Make 48 of each.

10. Lay out four light pink prairie-point units, five dark pink prairie-point units, six burgundy prairie-point units, and three medium pink print 2⅞" triangles as shown. Sew into rows and press. Sew the rows together and press. Make four units.

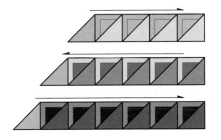

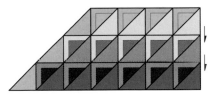

Make 4.

11. Sew a beige 6⅞" triangle to the edge of each unit from step 10. Press the seam allowances toward the beige triangle.

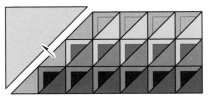

Make 4.

12. Lay out eight light pink prairie-point units, seven dark pink prairie-point units, six burgundy prairie-point units, three medium pink print 2⅞" triangles, and one each of medium pink, light green, and medium green 2½" squares as shown. Sew into rows, and press. Sew the rows together and press. Make four units.

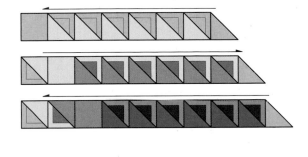

Make 4.

13. Sew a beige 6⅞" triangle to the edge of each unit from step 12. Press the seam allowances toward the beige triangle.

Make 4.

14. Lay out a unit from step 8, a unit from step 11, and a unit from step 13 as shown. Sew the units together. Press the seam allowances toward the tree-trunk unit. Make four blocks.

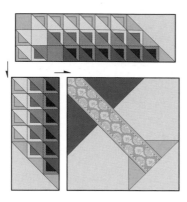

Make 4.

ASSEMBLING THE QUILT TOP

1. Referring to the assembly diagram, lay out the blocks, the beige 20½" square, and the beige setting triangles into diagonal rows. Sew the diagonal rows and press toward the beige fabric. Sew the rows together to complete the quilt top.

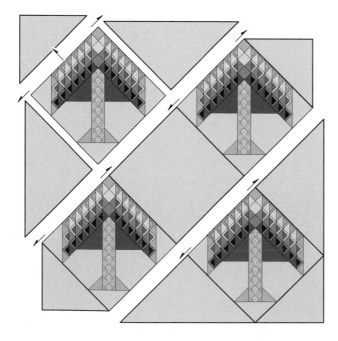

2. Referring to "Borders" (page 74), apply the border-print strips to the quilt top using the mitered-corners technique.
3. Sew two light green 3" x 42" strips together, end to end, to make one long strip. Make four strips.
4. Apply the light green border strips to the quilt top using the mitered-corners technique.

QUILTING AND BINDING

1. Refer to "Preparing the Quilt Sandwich" (page 76) to layer the quilt top, batting, and backing.
2. Hand or machine quilt as desired.
3. Referring to "Binding" (page 77), use the 2"-wide light green strips to bind the edges of your quilt.

QUILTING SUGGESTION

This quilt has a large amount of space for elaborate quilting. I created an original design for the center quilting block. If you'd like to use it, you can download the pattern at www.theniftyneedle.com. I quilted the same feathers and flowers in the setting and corner triangles and echo quilted and meandered around them. I also outline quilted around the block patches. I recommend using wool batting for this project; it really makes the quilting stand out.

STAR-STUDDED SAMPLER

I truly enjoyed making this quilt. With nine different blocks, there's no opportunity for boredom. In fact, I'd get so excited about how one block turned out that I could hardly wait to make the next one. I love the way all the blocks work together seamlessly and how the setting has stars within stars, culminating in one large star.

Finished quilt: 74½" x 74½"
Finished blocks: 12" x 12"

MATERIALS

Yardage is based on 42"-wide fabric.

- ► 2⅓ yards of brown floral print for blocks and outer border
- ► 2 yards of brown tone-on-tone print for blocks, middle border, and binding
- ► 1⅓ yards of pink tone-on-tone print for blocks and prairie points
- ► 1⅛ yards of beige floral print for blocks
- ► 1⅛ yards of dark pink tone-on-tone print for blocks, prairie points, and inner border
- ► ⅞ yard of large-scale floral print for setting blocks
- ► ⅝ yard of beige tone-on-tone fabric for blocks
- ► 5 yards of fabric for backing
- ► 82" x 82" piece of batting
- ► 13½ yards of ¼"-wide Steam-A-Seam 2
- ► Placement guide (see page 9)

> **LABEL PIECES TO STAY ORGANIZED**
>
> I suggest that you take a moment to label all your pieces as you cut them. I put sticky notes on each set of patches to indicate the cut size. This was extremely helpful when piecing the blocks: I could pull from the pile that said 5¼" triangles, or the one that said 4⅞" triangles, and so on. This simple step saved me lots of time later.

CUTTING

All measurements include ¼"-wide seam allowances.

From the pink tone-on-tone print, cut:
- 1 strip, 5¼" x 42"; crosscut into:
    3 squares, 5¼" x 5¼"; cut each square into quarters diagonally to yield 12 triangles
    2 squares, 4¼" x 4¼"; cut each square into quarters diagonally to yield 8 triangles

- 2 strips, 4" x 42"; crosscut into:
    12 squares, 4" x 4"
    4 squares, 3⅞" x 3⅞"; cut 2 squares in half diagonally to yield 4 triangles

- 3 strips, 3¾" x 42"; crosscut into 24 squares, 3¾" x 3¾"

Continued on page 61

Pieced and quilted by Karen Sievert

STAR-STUDDED SAMPLER

- 2 strips, 3⅝" x 42"; crosscut into:
  12 squares, 3⅝" x 3⅝"; cut each square in half diagonally to yield 24 triangles
  2 squares, 3½" x 3½"
  2 squares, 3¼" x 3¼"; cut each square into quarters diagonally to yield 8 triangles.
- 1 strip, 2¾" x 42"; crosscut into 8 squares, 2¾" x 2¾"
- 1 strip, 2⅜" x 42"; crosscut into 8 squares, 2⅜" x 2⅜". Cut each square in half diagonally to yield 16 triangles.
- 1 strip, 2" x 42"; crosscut into 16 squares, 2" x 2"
- 1 square, 4¾" x 4¾"
- 6 squares, 3" x 3"; cut each square in half diagonally to yield 12 triangles

From the dark pink tone-on-tone print, cut:
- 1 strip, 5¼" x 42"; crosscut into 7 squares, 5¼" x 5¼". Cut each square into quarters diagonally to yield 28 triangles.
- 1 strip, 4¼" x 42"; crosscut into:
  4 squares, 4¼" x 4¼", cut each square into quarters diagonally to yield 16 triangles
  4 squares, 4" x 4"
- 2 strips, 3⅞" x 42"; crosscut into:
  18 squares, 3⅞" x 3⅞; cut 8 of the squares in half diagonally to yield 16 triangles
  2 squares, 3½" x 3½"
- 1 strip, 3" x 42"; crosscut into:
  2 squares, 3" x 3"; cut each square in half diagonally to yield 4 triangles
  12 squares, 2¾" x 2¾"
- 1 strip, 2⅜" x 42"; crosscut into 8 squares, 2⅜" x 2⅜". Cut each square in half diagonally to yield 16 triangles.
- 1 strip, 2" x 42"; crosscut into 16 squares, 2" x 2"
- 8 strips, 1" x 42"

From the beige tone-on-tone print, cut:
- 1 strip, 7¼" x 42"; crosscut into:
  2 squares, 7¼" x 7¼"; cut each square into quarters diagonally to yield 8 triangles
  2 squares, 5¼" x 5¼"; cut each square into quarters diagonally to yield 8 triangles
- 1 strip, 4⅞" x 42"; crosscut into:
  4 squares, 4⅞" x 4⅞"
  2 squares, 2⅞" x 2⅞"; cut each square in half diagonally to yield 4 triangles.
- 1 strip, 3⅞" x 42"; crosscut into:
  6 squares, 3⅞" x 3⅞"; cut 2 of the squares in half diagonally to yield 4 triangles
  4 squares, 2⅝" x 2⅝"
- 1 strip, 2½" x 42"; crosscut into:
  4 squares, 2½" x 2½"
  4 squares, 1⅞" x 1⅞"

From the beige floral print, cut:
- 2 strips, 12⅞" x 42"; crosscut into:
  4 squares, 12⅞" x 12⅞"
  4 rectangles, 6½" x 12½"
- 1 strip, 6⅞" x 42"; crosscut into 4 squares, 6⅞" x 6⅞". Cut each square in half diagonally to yield 8 triangles.
- 1 strip, 4⅞" x 42"; crosscut into:
  4 squares, 4⅞" x 4⅞"
  4 squares, 3⅞" x 3⅞"

From the brown floral print, cut:
- 1 strip, 12½" x 42"; crosscut into 4 rectangles, 6½" x 12½"
- 1 strip, 6⅞" x 42"; crosscut into:
  4 squares, 6⅞" x 6⅞"; cut each square in half diagonally to yield 8 triangles
  1 square, 5¼" x 5¼"; cut into quarters diagonally to yield 4 triangles
- 1 strip, 4¾" x 42"; crosscut into:
  2 squares, 4¾" x 4¾"
  3 squares, 4½" x 4½"
- 1 strip, 3⅝" x 42"; crosscut into:
  6 squares, 3⅝" x 3⅝"; cut each square in half diagonally to yield 12 triangles
  1 square, 3⅜" x 3⅜"
- 8 strips, 6" x 42"

Continued on page 62

From the brown tone-on-tone print, cut:

- 1 square, 13¼" x 13¼"; cut in quarter diagonally to yield 4 triangles

- 2 squares, 5¼" x 5¼"; cut each square into quarters diagonally to yield 8 triangles

- 1 strip, 4½" x 42"; crosscut into 8 squares, 4½" x 4½"

- 1 strip, 4¼" x 42"; crosscut into 8 squares, 4¼" x 4¼". Cut each square into quarters diagonally to yield 32 triangles.

- 2 strips, 3⅝" x 42"; crosscut into 12 squares, 3⅝" x 3⅝"

- 1 strip, 3½" x 42"; crosscut into 8 squares, 3½" x 3½"

- 1 strip, 2" x 42"; crosscut into 16 squares, 2" x 2"

- 8 strips, 1½" x 42"

- 8 strips, 2" x 42"

From the large-scale floral print, cut:

- 4 squares, 12⅞" x 12⅞"

PIECING BLOCK A

The name of this block is Swamp Angel.

1. Referring to "Making Prairie Points" (page 8), make eight prairie points from the pink 3¾" squares.

2. Referring to "Fusing Prairie Points" (page 9), fuse the eight pink prairie points to eight dark pink tone-on-tone 5¼" triangles.

3. Lay out two pink prairie-point units, one beige tone-on-tone 5¼" triangle, and one pink 5¼" triangle as shown. Sew into pairs. Press the seam allowances away from the prairie-point units. Sew the pairs together to form an hourglass unit. Press the seam allowances open. Make four units. Trim the dog-ears.

Make 4.

4. Draw a diagonal line on the wrong side of two beige floral 4⅞" squares. With the marked squares on top, layer each square with a beige tone-on-tone 4⅞" square, right sides together. Stitch ¼" away from both sides of the drawn line. Cut the squares apart on the line to yield four half-square-triangle units. Press the seam allowances toward the beige floral triangles. Trim the dog-ears.

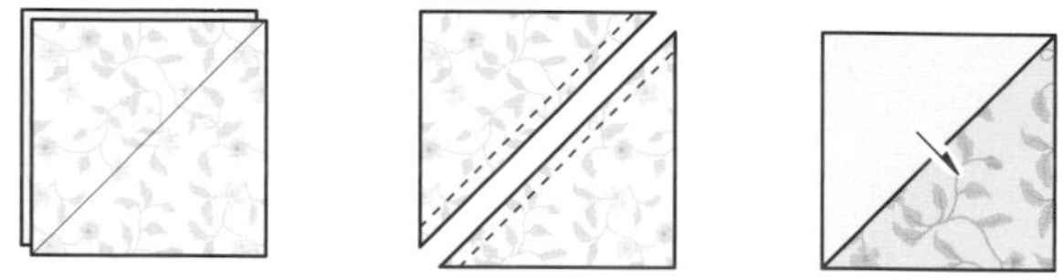

Make 4.

5. Lay out the four units from step 3, the four units from step 4, and one brown floral 4½" square into horizontal rows. Sew each row together. Press the seam allowances in the direction indicated. Sew the rows together, pressing toward the center.

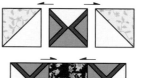 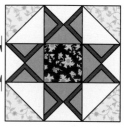

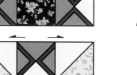

Block A

PIECING BLOCK B

Block B is a variation of the block called Mosaic.

1. Make eight prairie points from the pink 4" squares.

2. Fuse the eight pink prairie points to eight dark pink tone-on-tone 3⅞" triangles.

3. Draw a diagonal line on the wrong side of two beige floral 3⅞" squares. With the marked squares on top, layer each square with a beige tone-on-tone 3⅞" square, right sides together. Stitch ¼" away from both sides of the drawn line. Cut the squares apart on the line to yield four half-square-triangle units. Press the seam allowances toward the beige floral triangles. Trim the dog-ears.

Make 4.

4. Sew a prairie-point unit to one short side of a beige tone-on-tone 7¼" triangle. Press the seam allowances toward the beige triangle. Repeat on the other short side to make a flying-geese unit. Make four.

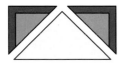 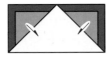

Make 4.

5. Sew dark pink tone-on-tone 3⅞" triangles to opposite sides of a brown floral 4¾" square. Press the seam allowances toward the dark pink triangles. Repeat for the remaining two sides of the square. Trim the dog-ears.

6. Lay out the four units from step 3, the four units from step 4, and the unit from step 5 into three horizontal rows as shown. Sew each row together, and press the seam allowances in the direction indicated. Sew the rows together to complete the block.

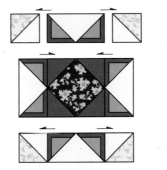 

Block B

PIECING BLOCK C

This is another variation of the Mosaic block.

1. Make four pink and four dark pink prairie points from 4" squares.
2. Fuse the pink prairie points to four dark pink tone-on-tone 3⅞" triangles and fuse the dark pink prairie points to four pink 3⅞" triangles.
3. Draw a diagonal line on the wrong side of two beige floral 3⅞" squares. With the marked squares on top, layer each square with a beige tone-on-tone 3⅞" square, right sides together. Stitch ¼" away from both sides of the drawn line. Cut the squares apart on the line to yield four half-square-triangle units. Press the seam allowances toward the beige floral triangles. Trim the dog-ears.

Make 4.

4. Draw a diagonal line on the wrong side of two pink 3⅞" squares. With the marked squares on top, layer each square with a dark pink tone-on-tone 3⅞" square, right sides together. Make four half-square-triangle

units as you did in step 3. Press the seam allowances toward the dark pink triangles.

Make 4.

5. Lay out the four units from step 4 as shown. Sew into pairs. Press the seam allowances open to reduce bulk. Sew the pairs together. Press the seam allowances open.

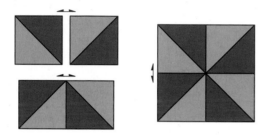

6. Sew a pink prairie-point unit to one short side of a beige tone-on-tone 7¼" triangle. Press the seam allowances toward the beige triangle. Repeat on the other short side with a dark pink prairie-point unit to make a flying-geese unit. Make four.

Make 4.

7. Lay out the four units from step 3, the four units from step 6, and the unit from step 5 into three horizontal rows. Sew each row together. Press the seam allowances in the direction

indicated. Sew the rows together to complete the block. Press the seam allowances toward the center.

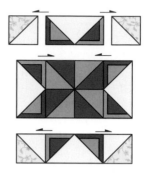 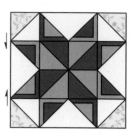

Block C

PIECING BLOCK D

Block D is a variation of the Swamp Angel block.

1. Make eight prairie points from pink 3¾" squares and fuse them to eight dark pink tone-on-tone 5¼" triangles.

2. Lay out two pink prairie-point units, one beige tone-on-tone 5¼" triangle, and one brown floral 5¼" triangle as shown. Sew into pairs. Press the seam allowances away from the prairie-point units. Sew the pairs together to form an hourglass unit. Press the seam allowances open. Make four units. Trim the dog-ears.

Make 4.

3. Draw a diagonal line on the wrong side of two beige floral 4⅞" squares. With the marked squares on top, layer each square with a beige tone-on-tone 4⅞" square, right sides together. Stitch ¼" away from both sides of the drawn line. Cut the squares apart on the

line to yield four half-square-triangle units. Press the seam allowances toward the beige floral triangles. Trim the dog-ears.

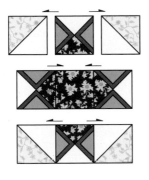

Make 4.

4. Lay out the four units from step 2, the four units from step 3, and a brown floral 4½" square into three horizontal rows. Sew each row together. Press the seam allowances in the direction indicated. Sew the rows together to complete the block. Press the seam allowances toward the center.

Block D

PIECING BLOCK E

This block is known as Rhode Island.

1. Make eight pink prairie points from 3¾" squares and fuse them to eight dark pink tone-on-tone 5¼" triangles.
2. Sew a pink 3¼" triangle to one short side of a beige tone-on-tone 2⅞" triangle. Press the seam allowances toward the pink triangle. Repeat on the other short side to make a flying-geese unit. Make four.

Make 4.

3. Lay out four beige tone-on-tone 1⅞" squares, the four units from step 2, and one brown floral 3⅜" square into three horizontal rows. Sew each row together. Press the seam allowances in the direction indicated. Sew the rows together. Press the seam allowances toward the center.

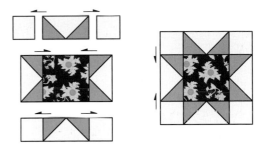

4. Lay out a brown tone-on-tone 4½" square and two prairie-point units as shown. Sew the prairie-point units to the sides of the square. Press the seam allowances toward the brown fabric. Make four units.

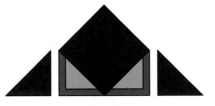

Make 4.

5. Sew brown tone-on-tone 5¼" triangles to opposite sides of the unit from step 4. Press the seam allowances toward the brown triangles. Make two units.

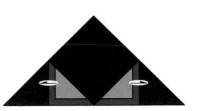

Make 2.

6. Sew the units from step 4 to opposite sides of the unit from step 3. Press the seam allowances open. Sew the two units from step 5 to the remaining two sides to complete the block. Press the seam allowances open.

Block E

PIECING BLOCK F

Block F is known as the Crown of Thorns block.

1. Make 12 prairie points from dark pink 2¾" squares and fuse them to 12 pink 3" triangles.
2. Sew a dark pink 3" triangle to a prairie-point unit along the long edge. Press the seam allowances toward the dark pink triangle. Make four units. Trim the dog-ears.

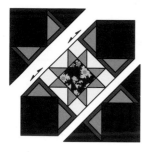

Make 4.

3. Sew a brown tone-on-tone 4¼" triangle to each side of the prairie point in the unit from step 2. Press the seam allowances toward the brown triangles. Make four units. Trim the dog-ears.

Make 4.

4. Sew a dark pink 4¼" triangle to a prairie-point unit along the short edges. Press the seam allowances toward the prairie-point unit. Make four units and four reversed as shown. Trim the dog-ears.

Make 4 of each.

5. Sew a unit and a reversed unit from step 4 to the sides of a unit from step 3. Press the seam allowances in the direction indicated. Make four units.

Make 4.

6. Sew beige tone-on-tone 3⅞" triangles to opposite sides of a brown floral 4¾" square. Press the seam allowances toward the beige triangles. Repeat for the remaining two sides of the square. Trim the dog-ears.

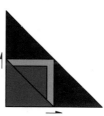

7. Lay out four brown tone-on-tone 3½"
   squares, the four units from step 5, and the
   unit from step 6 into three horizontal rows.
   Sew each row together. Press the seam
   allowances in the direction indicated. Sew the
   rows together to complete the block. Press.

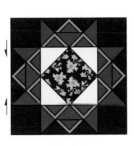

Block F

PIECING BLOCK G

I call this block Four Patch Variable Stars.

1. Make 16 dark pink and 16 pink prairie points
   from the 2" squares. Fuse the dark pink
   prairie points to 16 dark pink 2⅜" triangles
   and fuse the pink prairie points to 16 pink 2⅜"
   triangles.

2. Sew a dark pink prairie-point unit to each
   short side of a brown tone-on-tone 4¼"
   triangle. Press the seam allowances toward
   the brown triangle. Make eight units.

Make 8.

3. Lay out four brown tone-on-tone 2" squares,
   four units from step 2, and a dark pink 3½"
   square into three horizontal rows. Sew each
   row together. Press the seam allowances
   in the direction indicated. Sew the rows
   together. Press the seam allowances toward
   the center. Make two units.

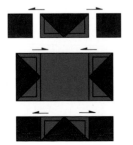

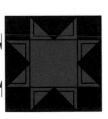

Make 2.

4. Repeat steps 2 and 3, substituting pink prairie
   points and pink 3½" squares to make two
   units in pink.

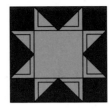

Make 2.

5. Lay out the dark pink and pink units into two
   horizontal rows, alternating the units from
   row to row. Sew together into pairs. Press
   the seam allowances open. Sew the rows
   together to complete the block. Press the
   seam allowances open.

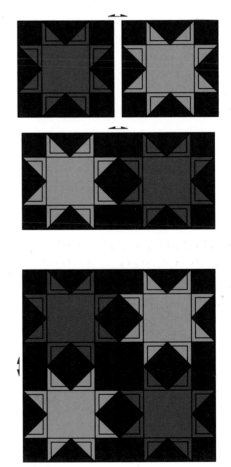

Block G

PIECING BLOCK H

Block H is called Ornate Star.

1. Make eight prairie points from the dark pink 3⅞" squares and fuse them to eight pink 5¼" triangles.
2. Lay out two prairie-point units, a dark pink 5¼" triangle, and a brown tone-on-tone 5¼" triangle as shown. Sew into pairs. Press the seam allowances away from the prairie-point units. Sew the pairs together to make an hourglass unit. Press the seam allowances open. Make four units. Trim the dog-ears.

Make 4.

3. Draw a diagonal line from corner to corner on the wrong side of four beige tone-on-tone 2½" squares. Layer a beige square on a brown tone-on-tone 4½" square, right sides together. Sew on the diagonal line. Trim ¼" away from the sewn line as shown. Press the seam allowances toward the beige triangle. Make four units.

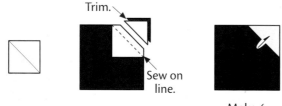

Make 4.

4. Lay out the four units from step 2, the four units from step 3, and a brown floral 4½" square into three horizontal rows. Sew each row together. Press the seam allowances

in the direction indicated. Sew the rows together to complete the block. Press the seam allowances open.

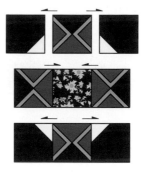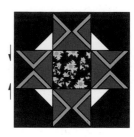

Block H

PIECING BLOCK I

This block is called Girl's Favorite.

1. Make eight prairie points from the pink 2¾" squares and fuse them to eight dark pink 4¼" triangles.
2. Sew a pink 4¼" triangle to the short side of a prairie-point unit as shown. Press the seam allowances toward the pink triangle. Make four and four reversed.

Make 4 of each.

3. Lay out one beige tone-on-tone 2⅝" square and two brown tone-on-tone 4¼" triangles. Sew the triangles to the sides of the square as shown. Press the seam allowances toward the brown triangles. Make four units. Trim the dog-ears.

Make 4.

STAR-STUDDED SAMPLER

4. Sew one unit and one reversed unit from step 2 to the short sides of the unit from step 3. Press the seam allowances toward the beige square. Make four units total. Trim the dog-ears.

Make 4.

5. Sew dark pink 3⅞" triangles to opposite sides of a pink 4¾" square. Press the seam allowances toward the dark pink triangles. Repeat for the remaining two sides of the square. Trim the dog-ears.

6. Lay out four units from step 4, the unit from step 5, and four brown tone-on-tone 3½" squares into three horizontal rows. Sew each row together. Press the seam allowances in the direction indicated. Sew the rows together to complete the block.

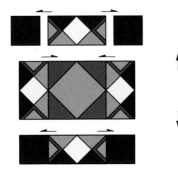
Block I

PIECING BLOCK J

To ensure a good fit with the other elements of the quilt, piece this triangular block with a scant ¼" seam allowance. When paired with a plain half-square triangle, the block is known as Birds in the Air.

1. Make 12 brown tone-on-tone prairie points from the 3⅝" squares and fuse them to 12 brown floral 3⅝" triangles.

2. Sew a pink 3⅝" triangle to a prairie-point unit along the long edge. Press the seam allowances toward the pink triangle. Make 12 units. Trim the dog-ears.

Make 12.

3. Lay out three units from step 2 and three pink 3⅝" triangles into three horizontal rows as shown. Sew the rows together. Press the seam allowances in the direction indicated. Sew the rows together. Make four blocks.

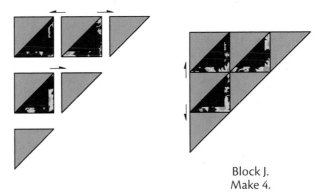
Block J.
Make 4.

ASSEMBLING THE QUILT TOP

1. Draw a diagonal line on the wrong side of the four beige floral 12⅞" squares. With the marked squares on top, layer each square with a large-scale floral print 12⅞" square, right sides together. Stitch ¼" away from both sides of the drawn line. Cut the squares apart on the line to yield eight half-square-triangle units. Press the seam allowances toward the brown floral triangles. Trim the dog-ears.

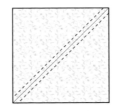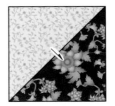
Make 8.

2. Lay out two beige floral 6⅞" triangles, and a block J. Sew the triangles to the short sides of block J. Press the seam allowances toward the beige floral triangles. Sew a beige floral 6½" x 12½" rectangle to the unit. Press the seam allowances toward the rectangle. Make four units.

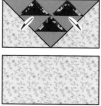
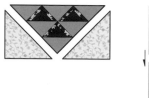

Make 4.

3. Lay out two brown floral 6⅞" triangles and one brown tone-on-tone 13¼" triangle. Sew the brown floral triangles to the short sides of the brown triangle. Press the seam allowances toward the brown floral triangles. Sew a brown floral 6½" x 12½" rectangle to the unit. Press the seam allowances toward the rectangle. Make four units. Trim the dog-ears.

Make 4.

4. Refer to the quilt assembly diagram on page 71 to lay out blocks A–I, the eight units from step 1, the four units from step 2, and the four units from step 3 into horizontal rows. Sew each row together. Press the seam allowances away from the blocks. Sew the rows together. Press the seam allowances all in one direction.

5. Sew two dark pink 1" strips together, end to end, to make one long strip. Make four long strips. Repeat with the brown tone-on-tone 1½" strips and the brown floral 6" strips.

6. Referring to "Borders" (page 74), apply the dark pink strips to the quilt top using the butted-corners method. Repeat to apply the brown tone-on-tone strips, and then the brown floral strips.

QUILTING AND BINDING

1. Refer to "Preparing the Quilt Sandwich" (page 76) to layer the quilt top, batting, and backing.
2. Hand or machine quilt as desired.
3. Referring to "Binding" (page 77), use the 2"-wide brown tone-on-tone strips to bind the edges of your quilt.

QUILTING SUGGESTION

Talk about possibilities—there are so many different directions you can go in quilting this quilt. Outlining your blocks to show off your incredible piecing skills is one option. In the beige floral background, you could stitch curling feathers, cross-hatching, meandering motifs, or even a combination of those ideas. However you decide to quilt it, my hope is that this quilt will be one that you treasure for years.

QUILTMAKING BASICS

In this section I'll cover basic techniques and provide helpful tips that will allow you to make quilts you are proud of. There may be other ways of accomplishing the same thing, but over the years, I have found these methods to be tried and true.

FABRIC CARE

This is a subject that I feel quite passionate about. I've learned the hard way, over and over again, that choosing the right fabric and preparing it carefully can make a big difference.

For instance, I *always* use 100% cotton fabric purchased from a quilt shop. As a new quilter, I thought I'd save money by buying less expensive fabrics, but found that it really ended up costing me more in the long run. Good quality, 100% cotton fabrics from quilt shops tend to hold up better.

I prewash all of my fabrics. The first thing I do when I come home with new fabric is to clip the corners at the selvage with pinking shears, and then put it in the washing machine. I've had three quilts that bled, one even after prewashing; I've learned that it's far better to discover that a fabric bleeds well before I've invested all my time and energy in a quilt. If you do have a finished quilt made with fabrics that haven't been prewashed, use a Color Catcher sheet (available at most grocery stores) in the washing machine when you launder it. The excess dye goes into the sheets instead of bleeding onto the quilt.

When I prewash my fabric, I never use fabric softener in the washing machine or in the dryer. I believe that one of the reasons quilters don't like to prewash is because they lose the sizing in the fabric. Using fabric softener contributes to this problem by making the fabric softer and more drapable, without that crisp feel it had on the bolt. To "resize" the fabric, I spray the washed and dried fabric with a good quality spray starch and let it rest for 15 minutes or so. This allows the starch to really soak into the fibers and eliminates flaking when pressing. I then press, using an iron on a high heat setting and using steam. One of the lessons I've learned is that a well-starched fabric is so much easier to work with when piecing; I experience less stretch on those pesky bias edges. It's a good idea to wash your quilt when completed to remove the starch.

ROTARY CUTTING

All of the patterns in this book are designed for rotary cutting. It's very important to ensure that you're cutting your pieces accurately. This is another area where I've learned my lessons the hard way. Believe me, if your cutting is off, your pieces won't match when you're ready to sew them. So, as the carpenter's adage goes, "Measure twice, cut once."

Strips

1. Fold your fabric in half lengthwise, wrong sides together, with the selvages aligned. Place the fabric on your cutting mat with the folded edge toward you. Align your ruler so that one of the horizontal lines is on the fold of the fabric and a small portion of the fabric is visible on the right-hand side of the ruler. Cut along the right edge of the ruler. This squares up the edge of the fabric.

2. Rotate either the fabric or the mat so that the newly straightened edge is to your left. To cut strips, place the ruler so that the desired measurement on your ruler is aligned with the cut edge of the fabric. For instance, to cut 3"-wide strips, align the 3" vertical line of the ruler with the cut edge of the fabric. Once

you're sure you have your ruler accurately placed, hold it stable with your noncutting hand and cut along the right edge of the ruler.

Squares and Rectangles

1. To cut squares, first cut a strip to the required measurement. Square up one end of the strip by cutting a small piece from the right side.

2. Rotate the strip so that the squared-up edge is to your left. Align the desired vertical measurement on the ruler with the cut edge of the fabric. Align a horizontal line of your ruler or the edge of the ruler with the bottom of the strip. Cut along the right edge of the ruler. To cut 3" squares from a 3" strip, align the 3" vertical line of the ruler with the cut edge of the fabric. To cut 3" x 4" rectangles from a 3" strip, align the 4" vertical line of your ruler with the cut edge of the fabric. Cut on the right side of the ruler.

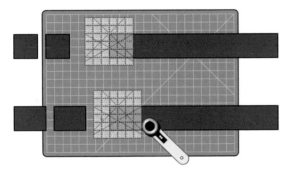

Half- and Quarter-Square Triangles

Most of the patterns in this book have half- and/ or quarter-square triangles in them. A half-square triangle is derived from a square that's been cut once diagonally and quarter-square triangles are derived from a square that's been cut twice diagonally.

Half-square triangles

Quarter-square triangles

PIECING

Once you've selected your fabric, washed it, pressed it, and cut it, it's time to sew. I can't stress enough the importance of an accurate ¼" seam. I once switched sewing machines in the middle of a project and couldn't figure out why my pieces weren't fitting together properly. Turns out the ¼" presser foot on one machine was actually ⅜". Well, ⅛" here and ⅛" there turns into ¼" off. And the problem keeps multiplying. So now, I *always* test my ¼" seam by sewing two measured scraps of fabric together, and then measuring the pieced unit.

Most sewing machines either come with a ¼" presser foot or the needle position can be moved to the left or right so that you can align the edge of the fabric with the edge of the presser foot and be able to stitch a ¼"-wide seam. If your machine has either of these options, sew a ¼" test seam and measure to make sure it truly is ¼".

If your sewing machine doesn't have a ¼" foot and if you can't adjust the needle position, mark a ¼" seam with masking tape. Do this by lowering the needle onto the ¼" mark of a ruler. Then place a strip of masking tape on the bed of the machine to mark the seam allowance, aligning the tape exactly next to the ruler. This gives you a guide for aligning the fabric as you sew.

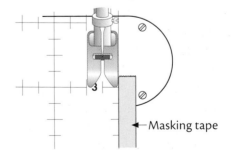

Masking tape

PINNING

Pinning is up to you, but keep in mind that pinning holds your patches and seams in place, countering the sewing machine's natural tendency to pull the top and bottom pieces through at different rates. I really hate seam ripping, and most of the time, if I have to do it, it's because I tried to save time by not pinning.

I always pin abutting seams and points using silk pins *without* glass ball heads. These pins glide easily in and out of the fabric and the lack of a ball head creates less distortion of the fabric as you're sewing. These pins are made by Collins and are available at quilt shops and online.

PRESSING—NOT IRONING

Pressing is another basic function of quilting, and again, experience has taught me a valuable lesson: there's a difference between pressing and ironing. Ironing is gliding the iron from side to side and top to bottom over the fabric. This is great for clothing but less than ideal for patchwork because it can distort seams and bias edges. I use a hot iron, usually *without* steam. Steam can distort bias edges quicker than anything else. When pressing, the heat of the iron, not the motion, smoothes wrinkles and sets seams.

To set a seam, first press it unopened from the wrong side of the fabric. Then open the piece and press again from the right side. Press the seam allowances in the direction indicated in the pattern. I use a straight-up-and-down approach and try not to move the iron from side to side.

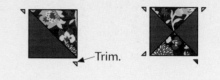
BORDERS

Like a beloved painting, borders are the frame for your beautiful patchwork. They encompass the quilt and tie it all together. Straight, mitered, pieced, or appliquéd, borders are the finishing statement.

Borders with Butted Corners

These straight borders are added to two opposite sides of the quilt first, and then to the remaining two sides.

1. Carefully press your quilt top and lay it flat. Measure the quilt top vertically through the center.

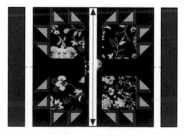

2. Cut border strips of the desired width to the measured length. When necessary, sew strips together, and then cut the strips to the exact length needed.

3. Place a pin in the center of the left and right sides of the quilt top and in the center of each of the side border strips.

4. With right sides together, match the center of the border strip to the center of the quilt top. Match the ends and pin. Add more pins between the center and ends, easing in the quilt top if necessary.

5. Using an accurate ¼" seam, sew the border strips to each side of the quilt top. Press the seam allowances toward the border strips.

6. Once again, lay the quilt top flat. Measure the quilt top through the center horizontally, including the borders you just added.

7. Repeat steps 2–5 to attach the top and bottom borders.

Borders with Mitered Corners

Border strips for mitered corners need to be cut longer than the measured width or length. I add double the width of the border strips plus 3" to the length of the border strips. For example, if I'm adding a 6"-wide border, I add 15" (6" + 6" + 3") to the strip length to ensure there is plenty of fabric to make the miter. Any excess fabric will be trimmed away after mitering.

I love border prints and I almost always miter the corners when I use them in a quilt. This requires a little extra planning, and sometimes extra fabric, but the end result is well worth it. Add an additional 10" or more to the length determined above if you're using a border print.

This will allow you to shift the border strips along the edge of the quilt and audition the motifs that will look best in the corners when joined at a 45° angle.

1. Measure through the quilt top vertically and horizontally. Cut border strips the width desired, and then the length as described above. Sew strips together if necessary.

2. Mark the center of each border strip with a pin. Divide the measured length of the quilt top in half and measure that distance away from the center pin in both directions on each side border strip. Place a pin at each of these points. For instance, if the quilt top measures 60", measure 30" from the center pin and mark that point at each end with a pin. Use the width of the quilt top and repeat the process for the top and bottom borders.

3. Pin the border to the quilt top, matching centers and raw edges. Make sure that the pinned edge meets the edge of the quilt top *exactly*.

4. *Very Important:* Begin sewing ¼" in from the pinned edge, backstitching to secure. Stop sewing when you are ¼" away from the opposite end and backstitch to secure. I usually mark the ¼" starting and stopping points with either a pin or pencil mark. Attach all four borders in this manner. Press the seam allowances toward the border.

5. Lay the quilt-top corner that you will be mitering right side up on an ironing board. Fold the border that is on top at a 45° angle and align it with the edges of the adjacent border. Press the fold to create a line on which to sew.

 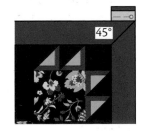

6. Fold the quilt top diagonally with right sides together to align the borders. Pin along the pressed line to keep the strips in place, and then sew along the pressed line starting at the previous ¼" stopping point and stitching out to the edge.

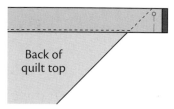

Back of quilt top

7. Open the borders to check the alignment of your seam. If you're happy with it, trim the excess fabric ¼" away from the sewing line. Press the seam allowances open.
8. Repeat steps 5–7 to miter the remaining three corners.

MULTIPLE MITERED BORDERS

You can simplify the addition of mitered borders by first piecing the inner border to the outer border and treating it as one border. If you're using a border print, decide where you want the miters to be in your border print first. Then center the inner-border strip on the outer-border strip and sew together before adding it to the quilt center.

PREPARING THE QUILT SANDWICH

After finishing your quilt top, measure the width and length. I like the batting and backing to be 4" larger all the way around the quilt top, so add 8" to the measured width and length. This is the size that your batting and backing should measure.

1. Lay the pressed backing, wrong side up, on a clean, flat surface. Use masking tape around the edges to prevent the backing from moving. The backing should be smooth and taut, but not stretched.
2. Center the batting over the backing and smooth out any wrinkles.
3. Center the quilt top, right side up, over the batting, again smoothing out any wrinkles. Make sure to check the back of your quilt top for loose threads before laying it over the batting. They can show through, especially under light fabrics.
4. Using size 1 or 2 curved basting safety pins, pin from the center out to the edges of the quilt top through all three layers, placing pins 3" to 4" apart. If you are hand quilting, baste the layers together with needle and thread.

QUILTING

The quilting stitches, whether done by hand or machine, serve a dual purpose. First, quilting holds the layers together, but second, and just as important, it adds to the finished beauty of the quilt. At times, trying to decide how to quilt a quilt is almost as hard as choosing the fabric for one!

I always look at my quilt tops and imagine them finished. I try to choose quilting designs that will have me saying, "I'm so glad I did that," rather than, "I wish I would have . . ." I strongly believe that the actual quilting has the ability to transform a quilt top and take it to a whole new level. The quilting design is as much a part of the design process as the color selection and piecing. Extra time and effort here will pay off well into the future. Today's quilts are tomorrow's heirlooms.

So what to do? First, take time to really look at your quilt top. Is it traditional? Then perhaps outline quilting the patchwork will showcase it best. Does it have large spaces for quilting motifs? Quilting designs such as a feathered wreath or floral bouquet can beautifully fill a space. Many patterns designed for blocks have matching border patterns to carry the quilting theme throughout. Would cross-hatching in the background add to the design? How about echo quilting? Only you can decide how you want the finished product to look.

Quilters are always asking me, "How do you quilt these quilts? Don't the prairie points get in the way?" I always respond by saying, "I quilt them just like any other quilt!" However, I should say that I never use allover patterns for these quilts. In fact, I rarely ever use them because I'm such a firm believer that "quilting makes the quilt."

For quilts with prairie points, I tend to subscribe to the KISS principle—Keep It Super Simple. For the most part, I outline quilt the patchwork, put motifs in large spaces, and try to carry that through to the border. Below are a few examples.

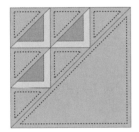
Outline quilting

Continuous curve

Meandering

In patterns where the prairie points are really large, I often quilt a small V in them. This keeps them from flopping open, yet still allows their dimension to shine through.

Upside-down "V" quilted
in prairie point

BINDING

I love putting the binding on a quilt simply because it means that my work is almost done. After attaching the binding to the front of the quilt by machine, I find it very rewarding and relaxing to do the final hand stitching. It gives me an opportunity to really see the beauty of what I've produced. So, let me share my favorite binding technique with you.

Preparing the Binding

I use 2"-wide strips for binding, resulting in a finished binding that's about ¼" wide. You may use a wider strip if you prefer.

1. Measure all the way around the quilt to determine the length of binding needed. Then add 10" to 12" for joining strips and turning corners. Let's say you have a quilt that is 60" x 72". The distance around the quilt comes to 264"; add 12" and you'll need 276" of binding.

2. Cut enough 2"-wide selvage-to-selvage strips from your binding fabric to equal the length you need from step 1. Cut both ends of your strips at a 45° angle, angling both ends in the same direction.

3. Measure ⅜" in from the tip of each strip and cut the tip off. Repeat on the opposite end of each strip. This will help you match the ends perfectly and you won't have to trim off any dog-ears.

Cut ⅜" tip from each end.

4. With right sides together, align and sew the ends together using a ¼" seam allowance. Press the seam allowances open.

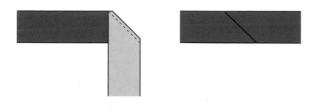

5. Fold the binding in half lengthwise, wrong sides together, and press.

Attaching the Binding

1. Align the raw edges of the binding with the raw edges of the quilt. Pin to the first corner, leaving the first 10" unpinned.

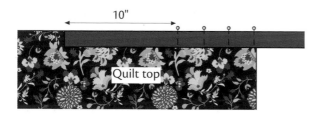

2. Using an accurate ¼" seam allowance, sew the binding to the quilt along the pinned edge. Stop ¼" from the corner and backstitch to secure.

3. Fold the binding *up* from the quilt at a 45° angle. Fold again, bringing the binding down toward the quilt and aligning the folded edge at the top with the raw edge of the quilt.

4. Pin the binding to the next side of the quilt. Start sewing ¼" from the top folded edge, and then backstitch to secure the stitching. Sew toward the next corner and stop ¼" from the end.

5. Repeat steps 3 and 4 at each corner.

6. After you've finished the fourth corner, stop sewing about 12" away from the beginning tail of the binding strip and remove the quilt from the machine.

7. Lay the quilt on a flat surface. Unfold the ends of the binding strip and overlap them, right sides together with the beginning tail on top. Mark the lower strip with a pencil where they overlap. Cut the marked strip ½" away from the line as shown.

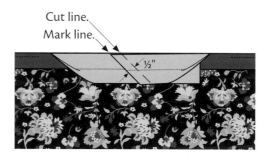

8. Fold the quilt out of the way and sew the beginning and ending binding strips together using a ¼" seam allowance, as you did when joining strips to make the binding.

9. Refold the binding, pin, and finish sewing the binding to the quilt.

10. Fold the binding over the raw edges of the quilt and hand sew it to the back using a blind stitch.

ABOUT THE AUTHOR

Karen Sievert began her quilting career in 1997 after being dragged into a quilt shop by her sister Barbara. Immediately hooked, Karen has been designing, teaching, and quilting ever since. It's a hobby that has truly become a passion for her. Like many other quilters, she gets a kick out of taking perfectly good fabric, cutting it up, and sewing it back together again.

What Karen loves most about quilting are the friendships she's made with other quilters. Her husband's military career has meant a lot of moves for her and her family, but she says, "No matter where I go, I'll meet other quilters, and voilà! New friendships are forged!"

An award-winning quilter and the author of *Better Together* (Martingale & Company, 2010), Karen shares her knowledge with other quilters through trunk shows, lectures, and workshops. She enjoys the inspiration and motivation she receives from traveling and seeing all the wonderful quilts that are out there.

Karen resides in Gainesville, Virginia, with her husband, Vince, and their three children, Wayne, Shannah, and Travis.

THERE'S MORE ONLINE!

Visit Karen at www.theniftyneedle.com for her acrylic placement guide and other notions, how-to videos, alternate layouts for the "Color, Color" quilts, and free downloadable quilt patterns. Visit www.martingale-pub.com for other great quilting books.

YOU MIGHT ENJOY THESE OTHER FINE TITLES FROM MARTINGALE & COMPANY

Our books are available at bookstores and your favorite craft, fabric, and yarn retailers.
Visit us at www.martingale-pub.com or contact us at:

1-800-426-3126
International: 1-425-483-3313
Fax: 1-425-486-7596
Email: Info@martingale-pub.com

Martingale®
& COMPANY

America's Best-Loved Craft & Hobby Books®
America's Best-Loved Knitting Books®

That Patchwork Place®

America's Best-Loved Quiting Books®